LINKING INDIA AND EASTERN NEIGHBOURS

Thank you for choosing a SAGE product!
If you have any comment, observation or feedback,
I would like to personally hear from you.

Please write to me at **contactceo@sagepub.in**

Vivek Mehra, Managing Director and CEO, SAGE India.

Bulk Sales

SAGE India offers special discounts
for purchase of books in bulk.
We also make available special imprints
and excerpts from our books on demand.

For orders and enquiries, write to us at

Marketing Department
SAGE Publications India Pvt Ltd
B1/I-1, Mohan Cooperative Industrial Area
Mathura Road, Post Bag 7
New Delhi 110044, India

E-mail us at **marketing@sagepub.in**

Subscribe to our mailing list

Write to **marketing@sagepub.in**

This book is also available as an e-book.

LINKING INDIA AND EASTERN NEIGHBOURS

Development in the Northeast and Borderlands

SAGE STUDIES ON INDIA'S NORTH EAST

Edited by
H. Srikanth
Munmun Majumdar

Los Angeles | London | New Delhi
Singapore | Washington DC | Melbourne

First published in 2021 by

SAGE Publications India Pvt Ltd
B1/I-1 Mohan Cooperative Industrial Area
Mathura Road, New Delhi 110 044, India
www.sagepub.in

SAGE Publications Inc
2455 Teller Road
Thousand Oaks, California 91320, USA

SAGE Publications Ltd
1 Oliver's Yard, 55 City Road
London EC1Y 1SP, United Kingdom

SAGE Publications Asia-Pacific Pte Ltd
18 Cross Street #10-10/11/12
China Square Central
Singapore 048423

Published by Vivek Mehra for SAGE Publications India Pvt Ltd. Typeset in 10.5/13 pt Bembo by AG Infographics, Delhi.

Library of Congress Control Number: 2021941443

ISBN: 978-93-91370-72-5 (HB)

SAGE Team: Amrita Dutta, Vandana Gupta and Anupama Krishnan

Contents

Part I: India's Relations with Neighbours: Politics of Development in NER

Part II: Border Crossing and Inter-Community Relations: Cooperation and Conflict

List of Illustrations

FIGURES

TABLES

List of Abbreviations

AA	Arakan Army
ADB	Asian Development Bank
AEP	Act East policy
AEZs	Agri Export Zones
AFD	Agence Française de Développement
AIIB	Asian Infrastructure Investment Bank
ASEAN	Association of Southeast Asian Nations
ASI	Archaeological Survey of India
ATS	Amphetamine-type stimulants
BADP	Border Areas Development Programme
BBIN	Bangladesh–Bhutan–India–Nepal Initiative
BCIM	Bangladesh–China–India–Myanmar
BGB	Border Guards Bangladesh
BIMSTEC	Bay of Bengal Initiative for Multi-sectoral Technical and Economic Cooperation
BRI	Belt and Road Initiative
BRICS	Brazil, Russia, India, China and South Africa
BRO	Border Roads Organisation
BRWS	The Burma Repatriates Welfare Society
BSF	Border Security Force
CBTE	Cross Border Trade of Electricity
CCGT	Combined-cycle gas turbine
CEDAW	Convention on the Elimination of All Forms of Discrimination Against Women
CEPA	Comprehensive Economic Partnership Agreement
CESPR	Centre for Environment, Social and Policy Research
CHRO	Chin Human Rights Organization
CNA	Chin National Army
CNF	Chin National Front
COMESA	Common Market for Eastern and Southern Africa

COVA	Confederation of Voluntary Associations
CPEC	China–Pakistan Economic Corridor
CRC	Convention on the Rights of the Child
CSO	Central Statistical Office
DFTP	Duty Free Tariff Preference
DPR	Detailed Project Report
EAPs	Externally aided projects
EDF	Export Development Fund
ERIA	Economic Research Institute of ASEAN and East Asia
FDI	Foreign direct investment
FMR	Free Movement Regime
GAIL	Gas Authority of India Limited
GATT	General Agreement on Tariffs and Trade
GDP	Gross Development Product
GiZ	Gesellschaft für Internationale Zusammenarbeit
GMS	Greater Mekong Subregion
GNH	Gross National Happiness
GQ	Golden Quadrilateral
GST	Goods and Services Tax
HRW	Human Rights Watch
ICCPR	International Covenant on Civil and Political Rights
ICESCR	International Covenant on Economic, Social and Cultural Rights
ICJ	International Commission of Jurists
ICP	Integrated Check Post
IFAD	International Fund for Agricultural Development
IIIT	International Institute of Information Technology
IMBAX	India–Myanmar Bilateral Military Exercises
IMBF	Indo-Myanmar Border Force
IMS-GT	Indonesia–Malaysia–Singapore Growth Triangle
IMTTH	India–Myanmar–Thailand Trilateral Highway
INA	Indian National Army
IOM	International Organization for Migration
JBWG	Joint Boundary Working Group
JICA	Japan International Cooperation Agency
JICF	Japan–India Coordination Forum
KfW	Kreditanstalt für Wiederaufbau (Credit Institute for Reconstruction)

KMMTP	Kaladan Multi-Modal Transit Transport Project
KNO	Kuki National Organization
LEP	Look East Policy
LSD	Lysergic acid diethylamide
MDMA	3,4-Methylenedioxy-N-hydroxy-N-methylamphetamine
MDoNER	Ministry of Development of North Eastern Region
MEA	Ministry of External Affairs
MHA	Ministry of Home Affairs
MHIP	Mizo Hmeichhe Insuihkhawm Pawl
MIEC	Mekong–India Economic Corridor
MJAC	Mizo Joint Action Committee
MoRTH	Ministry of Road Transport and Highways
MoU	Memorandum of Understanding
MUP	Mizo Upa Pawl
MZP	Mizo Zirlai Pawl
NAB	Narcotics and Affairs of Border
NCB	Narcotics Control Bureau
NDA	National Democratic Alliance
NDDB	National Dairy Development Board
NEC	North Eastern Council
NEDFi	North Eastern Development Finance Corporation Ltd
NEEPCO	North Eastern Electric Power Corporation Limited
NEHHDC	North Eastern Handicrafts and Handlooms Development Corporation Limited
NEIIP	North East Industrial and Investment Promotion Policy
NER	North Eastern Region
NERAMAC	North Eastern Regional Agricultural Marketing Corporation Limited
NERSDS	North East Road Sector Development Scheme
NESRIP	North Eastern State Roads Investment Programme
NHAI	National Highways Authority of India
NHDP	National Highways Development Project
NHIDCL	National Highways and Infrastructure Development Corporation Limited
NITI	National Institution for Transforming India

NRC	National Registration Card
NSCN	National Socialist Council of Nagaland
NSCN-IM	National Socialist Council of Nagaland (Isak–Muivah)
NS-EW	North–South and East–West Corridor
NSSO	National Sample Survey Office
OBOR	One Belt One Road
OECD	Organisation for Economic Co-operation and Development
ONGC	Oil and Natural Gas Corporation
PH	Pseudoephedrine
PMGSY	Pradhan Mantri Gram Sadak Yojana
POK	Pakistan Occupied Kashmir
PRC	People's Republic of China
PWD	Public Works Department
RSDP	Road Sector Development Programme
SAARC	South Asian Association for Regional Cooperation
SAFTA	South Asian Free Trade Area
SAGAR	Security and Growth for All in the Region
SAGQ	South Asia Growth Quadrangle
SARDP-NE	Special Accelerated Road Development Programme in the North East
SASEC	South Asia Subregional Economic Cooperation
SDP	State Domestic Product
SoOs	Suspensions of Operations
TCS	Tata Consultancy Services
TFA	Trade Facilitation Agreement
TIDC	Tripura Industrial Development Corporation Limited
UN	United Nations
UNCLOS	United Nations Convention on the Law of the Sea
UNDP	United Nations Development Programme
UNHCR	United Nations High Commissioner for Refugees
UNODC	United Nations Office on Drugs and Crime
UWSA	United Wa State Army
VPDC	Village Peace and Developmental Council
WTO	World Trade Organization
YMA	Young Mizo Association

Acknowledgements

The change and uncertainty that followed the collapse of the Soviet Union compelled Government of India to come out with the Look East policy, now rechristened as the Act East policy. In recent decades, Government of India has joined different sub-regional development initiatives involving China and Association of Southeast Asian Nations (ASEAN) and Indo-Pacific countries. Different trade and connectivity projects underway link India's North Eastern Region with its eastern neighbours. The conversation for this edited volume, *Linking India and Eastern Neighbours: Development in the Northeast and Borderlands*, started at an international seminar titled 'Northeast India and Its International Neighbors: New Directions', organized by the Department of Political Science, North-Eastern Hill University (NEHU), in 2018 as part of its University Grants Commission (UGC) Special Assistance Programme (SAP). The UGC-SAP enabled the department to undertake research projects and organize seminars and workshops every year. We are thankful to UGC for granting SAP to our department.

The department is obliged to the university administration for providing logistic support to all SAP-related activities. Apart from the UGC and the University, we received financial assistance for organizing seminars and workshops from the Indian Council of Social Science Research and the ICSSR North Eastern Regional Centre (ICSSR–NERC). We are thankful to these institutions that sponsored our programmes in the university.

The annual UGC-SAP seminars/workshops used to be meeting places where the faculty and research scholars from the university met scholars, academicians and journalists from other north-eastern states and also from other parts of India. We sincerely acknowledge the participation and contributions of the academicians, research scholars and journalists who took an active part in all activities.

We are indebted to senior scholars from India and from the countries neighbouring Northeast India for their contributions to the volume. All chapters included in the volume went through rigorous editorial scrutiny and revisions to meet the standard set by the editorial team and the publishers. We appreciate the patience and hard work that the authors have put in to give shape to the volume.

We also acknowledge the support and cooperation of all our colleagues, research scholars and supporting staff of the department. Professor K. Debbarma, Deputy Coordinator, and Dr Mary Magdalyne Kurbah, SAP Project Fellow, need special mention.

We are happy that SAGE Publications (India) has come forward to bring out the volume. Ms Amrita Dutta, the commissioning editor, guided and assisted us from the day we approached the publisher with our book proposal. We are deeply obliged to Amrita Dutta, Vandana Gupta and other members of the editorial team of SAGE for their efforts in bringing out the volume.

The book, breaking out of traditional paradigmatic ways of thinking, has made efforts to capture complex processes of change taking place in India's North Eastern Region and its borderlands. We hope our work will be a useful guide for students, policymakers and the media.

Introduction

H. Srikanth and Munmun Majumdar

Over the years, border studies has emerged as an insightful multidisciplinary field, drawing inputs from a host of disciplines such as geography, anthropology, political science, international relations, economics, history, native studies and post-colonial scholarship. What started in the 1970s as investigations of the complex processes at work in the borderlands between Mexico and the United States, and between Canada and the United States (Hämäläinen and Truett 2011) gradually moved beyond North America to focus on the study of borders in Europe, and then to the countries in Africa and Asia (Parker & Vaughan-Williams 2014; Rytövuori-Apunen and Marlin-Bennett 2016). Not confining the inquiry to the disputes over the territorial boundaries of nation states, border studies began focusing on social, cultural, economic and political processes at work in the borderlands. Various border-related issues, such as migration, refugee problem, trafficking, citizenship, cultural conflicts, ethnic movements, social identities, security, economic development and political sovereignty, are debated and discussed as part of border studies (Diener 2012; Wastl-Walter 2019; Wilson and Donnan 2012).

The salience of the national border has been acknowledged by scholars working on the issues related to nationalism and nation states. When the Soviet Union collapsed and the Cold War ended, many scholars felt that the forces of capitalist globalization would make the borders irrelevant sooner than later (Ceglowski 1998; Cox 2004). The World Trade Organization (WTO) advocated the opening of borders for easy movement of capital, technologies and commodities, and global financial institutions insisted on structural changes in the national economies to facilitate the free flow of capital and goods. These changes, expected to benefit the national economies, have in practice enhanced the power and influence of the non-state actors to whom borders make little sense (Kitching 2001; Wonders 2007). The increasing global inequalities and the unabated incidences of ethnic conflicts around the globe led to organized attempts by illegal migrants and refugees to cross national borders in search of a better life (de Wenden 2020). Further, the United States-led war on terror and the rapid growth of transnational crimes de-territorialized borders and compelled nation states to strengthen their border security arrangements and introduce new modes of surveillance to deal with illegal border crossing (Andreas 2002; William and Vlassis 2005).

Scholars have looked at bordering processes from different per-spectives—classical, neoclassical, Marxist, feminist and post-colonial. Border studies realize that there cannot be one comprehensive theory of borderlands, as borderlands may differ from one border to another; even within the same country, there can be different borderlands and border processes. Each borderland is unique. One can conduct as many studies as there are borderlands. As the number of states has increased, the studies on border issues have also multiplied.

These varied accounts of bordering processes are no more confined to Western countries. In recent decades, one can see the growing interest in understanding the border processes at work in developing countries. There are quite a few scholarly studies on many disputes and conflicts over territorial boundaries in different Asian and African countries (Farrelly 2012; van Schendel 2005). In India, there is no dearth of studies on India's border disputes with such countries as Pakistan, China and Nepal (Arpi 2013; Maxwell 2015; Nayak 2020; Ranjan 2018). Most studies look at borders from the perspective of nation states. One can see a few media outlets reporting the ground situation near the border during wars. However, serious academic stud-ies on what happens in the borderlands where the wars are fought are very few. Further, wars and security concerns are not the only reasons to study borders and border processes. Academia needs to capture the multiple manifestations of the border processes at work in developing countries. Our knowledge of how people live in borderlands, what kinds of problems they encounter and how governments have been responding to the needs and aspirations of border people is still limited. In recent years, some scholars have come out with interesting studies on select borderlands, including the borderlands of the North Eastern region (NER) (Gogoi 2020; van Schendel 2005; Saikia & Chaudhury 2019; Mukherjee 2019). However, many more insightful studies on the borderlands of India's NER and their links with the countries neighbouring them are required.

HISTORY OF DELINEATION OF INDIA'S BORDERS

India has been a cultural and civilizational entity for thousands of years. However, the notion of India as a single political entity extending from Kashmir to Kanyakumari, and from Saurashtra to Silchar (or from Okha to Wokha), is a recent phenomenon that has its roots in the colonial and post-colonial history of the Indian subcontinent. Prior to colonization by the British, India was divided into different kingdoms and empires whose frontiers changed constantly, depending on the strength of the king's or the emperor's sword (Lessman 2015). One could see variations in territorial country within the same dynasties. The extent of territory held by Ashoka differed from that by Chandra Gupta Maurya. Similarly, the frontiers of the empires of Akbar and Aurangzeb were different. Some kings and emperors who ruled different parts of the subcontinent had their frontiers extended up to Kandahar (Afghanistan) in the West and Malay (South East Asia) in the East. As the boundaries of the kingdoms/empires constantly changed, it is difficult to distinguish between a native and a foreigner. As even Karl Marx observed, most people in the subcontinent lived in autonomous villages, and the changing boundaries of the kingdoms and empires had little impact on their lifeworld. However, all this changed with the establishment of the colonial rule by the British (Marx 1853). It was the British that brought the Indian subcontinent, then under different chiefs, kings and emperors, under one political and administrative rule. In their own colonial interest, the British took decisions on which political territories would become part of British India and which would be excluded from their direct rule, how far their colonial empire would extend, who would be their international neighbours, what treaties would be concluded with the native kings and with the powers that ruled the neighbouring territories and which territories needed to be kept as a buffer, partially excluded, excluded or left non-administered. It was the common subjection of people belonging to different communities and territories to direct or indirect rule by the British which stirred the nationalist sentiments and gave birth to sovereign India, that is, Bharat.

However, because of contradictions within the anti-imperialist movement, India did not inherit all the territories under the control of British India. Owing to the rebellious nature of the Afghans, the British found it difficult to subdue them and make them part of British India (Leake 2016). Although Burma was made part of British India after 1890, the natives of Burma could not identify themselves as Indians, and their demand for autonomy compelled the British to take a decision to separate Burma from British India in 1937 (Kent 2017). Failure to resolve communal tensions between Hindus and Muslims forced Indian nationalist leaders to agree to the partition of British India, paving the way for the birth of two sovereign states—India and Pakistan. In Sylhet district in Assam, a referendum was conducted to decide on whether the district would be part of India or would become part of East Pakistan (Bhattacharjee 1990). Immediately after India became independent, the Indian nationalist leaders took initiatives to bring all contiguous princely states and excluded or non-administered tribal territories under India's control. By the time India adopted its own Constitution and became a republic in 1950, what territories would be part of India and how far India's rule would extend became clear.

Just as it took years to become clear on what makes up the territorial boundaries of modern India, the nature and dynamics of India's immediate neighbourhood also developed in the course of time. The British viewed Nepal, Bhutan, Sikkim and Tibet as buffer states that would keep China and Russia at a safe distance from British India. Immediately after independence, India entered into friendship treaties with its northern neighbours, namely Nepal, Bhutan (1949) and Sikkim (1950). With its takeover of Tibet in the mid-1950s, China emerged as an important neighbour on India's north and north-eastern side. The McMahon Line, which India had assumed as its border with China, became contentious, as the Chinese contested the legality of the line. In contrast, India had no serious issue with Burma as far as the Indo-Burmese border was concerned. The leaders of India and Burma formalized and approved the boundary demarcation between India and Burma. The partition of British India and the birth of Pakistan have made Afghanistan a distant neighbour. Two important developments in the 1970s slightly altered the border situation in the NER. The conflict between East Pakistan and West Pakistan, and India's support to the freedom movement in East Pakistan, resulted in the bifurcation of Pakistan and the birth of Bangladesh in 1971 as a new neighbour on the eastern and north-eastern front. Further, as a follow-up to a contentious referendum, Sikkim, which had till then been a protectorate state, became an integral part of the Indian Union in 1975 (Sethi 1978; Sidhu 2018). Sri Lanka and Mauritius, separated by the Indian Ocean, remained India's neighbours on the southern front. By the mid-1970s, India's neighbourhood had evolved.

INDIA'S LAND BORDER WITH NEIGHBOURING COUNTRIES

Apart from 7,516.6 km long coastline, which includes island territories, India shares land borders of length around 15,000 km with its neighbouring countries. Seven states, namely Maharashtra, Goa, Karnataka, Kerala, Tamil Nadu, Andhra Pradesh and Orissa share only coastlines and no land borders. Two states—Gujarat and West Bengal—share both land borders and coastlines. Five states (Haryana, Madhya Pradesh, Jharkhand, Telangana and Chhattisgarh) and two union territories (UTs) (Chandigarh and Delhi) have neither land borders nor coastlines connecting them with neighbouring countries. According to the Ministry of Home Affairs, Government of India, the country shares land borders with Pakistan (3,323 km), China (3,488 km), Bangladesh (4,096.7 km), Nepal (1,751 km), Bhutan (699 km), Myanmar (1,643 km) and Afghanistan (106 km) (Ministry of Home Affairs 2019). As India is not in effective control of Pakistan Occupied Kashmir (POK), one may contest its claim of a 1,597-km-long border with China and a 106-km-long border with Afghanistan. Seventeen states and two UTs in India share land borders with one or more neighbouring countries. Whereas the states in western India share a border with Pakistan, some northern states border China and Nepal. All states in India's NER have borders with one or more neighbouring countries barring Pakistan. Table I.1 shows the states and union territories that share land borders with neighbouring countries.

Table I.1 *States and Union Territories Bordering India's Neighbours*

Neighbouring Country	States Sharing Land Borders (Length in km)	Union Territories Sharing Land Borders (Length in km)	Total Length of the Border Shared with a Neighbouring Country (Length in km)
Pakistan	Gujarat (506), Rajasthan (1,170), Punjab (425)	Jammu and Kashmir (1,222)	3,323
China	Himachal Pradesh (200), Uttarakhand (345), Sikkim (220) and Arunachal Pradesh (1,126)	Ladakh (1,597)	3,488
Nepal	Uttarakhand (275), Uttar Pradesh (551), Bihar (726), West Bengal (100) and Sikkim (99)		1,751
Myanmar	Mizoram (510), Manipur (398), Nagaland (215) and Arunachal Pradesh (520)		1,643
Bangladesh	West Bengal (2,216.7), Assam (263), Meghalaya (443), Tripura (856) and Mizoram (318)		4,096.7
Bhutan	Arunachal Pradesh (217), Sikkim (32), West Bengal (183) and Assam (267)		699

Source: Annual Report 2018–2019, Ministry of Home Affairs, Government of India.

INDIA'S NORTH-EASTERN BORDERS

India's NER has a unique history. The region has been the abode of different ethnic, racial, linguistic and cultural groups, most of which are unfamiliar to the citizens in the rest of India. During the pre-colonial period, some territories and communities in the region were no doubt influenced by religious and cultural traditions of mainland India. Still, there were several indigenous hill communities in the region who continued to practise their own native religious practices and traditions. Many scholars and social activists in the NER point out that the region was never under the occupation of any emperors and kings that ruled mainland India. Many Mongoloid communities settled in the NER locate their historical links to China and different Southeast Asian countries. There are many studies that show the economic vibrancy of the region during the pre-colonial and colonial era (Baruah 2003; Misra 2001). In their writings, several colonial scholars and administrators pointed out vibrant trade practices within the region and between the NER and the neighbouring Tibet, Bhutan and Bengal. The region was not primitive and backward.

However, the consolidation of the colonial rule compelled major changes in the region. The colonial administrative organization of the region into provinces, princely states and districts and into excluded and non-administered areas impacted the ethnic identities of the people in the region. The British authorities dictated the trajectory of development in the region, influencing all aspects of development, be it agriculture, plantations, mining, trade, employment, immigration, settlement or movement of the people. Apart from initiating economic policies and administrative changes in the region to serve their own colonial interests, the British pursued a frontier policy that had contradictory effects on the lives of the communities inhabiting the region. Their wars and friendship with the rulers of Nepal, Burma and Tibet and the conclusion of different treaties with them had long-term implications on the region. The Shimla Convention 1913–1914 that ratified the McMahon Line, the Government of India Act of 1935 that separated Burma from British India and the First and the Second World Wars that the British fought also had contradictory effects on the subsequent developments in the region. Finally, the decision to partition British India into India and Pakistan, and the referendum in the Sylhet district of East Bengal, compelled thousands of Hindu Bengalis to migrate to Assam and Tripura, influencing subsequently the politics of post-colonial Assam and Tripura (Bhattacharjee 2009; Ghoshal 2010–2011; Ramunny 1988).

DILEMMAS OF BORDERLANDS IN INDIA'S NORTH EASTERN REGION

Comparatively, for different historical reasons, there has been a considerable academic and policy focus on India's borders in its western and northern Fronts. In contrast, for long, India's NER that comprises eight states has not received adequate attention, although the region borders five countries—China, Bangladesh, Nepal, Bhutan and Myanmar. Because of its peripheral status and geographical location, policymakers did not give adequate attention to the development of the region till the mid-1980s. During the Sino-Indian War of 1962, the Chinese troops reached almost up to Tezpur, before they unilaterally declared ceasefire and withdrew to the Line of Actual Control (Panag 2018). The defeat at the hands of China compelled the Indian policymakers to view the region from a security-centric perspective. The breakout of ethnic insurgencies in different north-eastern States and the logistic support that some of them received from neighbouring countries also strengthened the security concerns in the region, leading to deployment of military and paramilitary forces to quell the militancy (Bhaumik 2009; Hazarika 1994; Nag 2002a; Sahni 2001; Upadhyay 2019). Other factors, such as the impact of the partition that disrupted the communities' organic links or dependency with the neighbouring territories and communities, economic neglect, growing unemployment, illegal immigration, refugee problem, trafficking of drugs and small arms have led to intensification of social and ethnic tensions in the region. Over the years, there has been considerable academic and policy literature on the various problems of the NER (Baruah 2020; Dutta and Choudhury 2017).

It was only after the Assam Movement (1979–1985), which influenced the politics of all north-eastern states, that the Indian government realized the need for economic intervention and border management. Since the mid-1980s, Government of India has been pumping crores of rupees into developmental activities in the region. The list of activities initiated by the central government for the development of the NER is provided on the Ministry of Development of North Eastern Region (MDoNER) website. Apart from activating institutions like the North Eastern Council, the central government has announced special packages and allocations from time to time for the development of infrastructure in the region. Special focus has been given to fencing of the India–Bangladesh border to prevent illegal migration (Saddiki 2017). For over a decade, the central government has been implementing the Border Areas Development Programme (BADP) in the border areas of different north-eastern states (Das 2010). While continuing the anti-insurgency operations, the government has also started peace talks with different ethnic militant groups in the region. It has established several universities and institutions of higher learning in the region over the last three decades. Politicians and bureaucrats have been talking of bridging the gulf between the NER and the rest of India and of bringing the people of the Northeast into the national mainstream. However, things did not move as expected. Each effort planned to break the region's isolation was countered by measures that restricted the connectivity, each new set of policies designed to boost the region's economies led to the patronage of regional elites in the region, and each new initiative to enhance participation from civil society was neutralized by the enactment of Black Acts and counter-insurgency operations. Such contradictory practices continued to exacerbate the grievances and problems in the region.

Meanwhile, the dramatic changes at the global level kick-started by the fall of the Soviet Union in 1990 and the decline of socialist ideology compelled Government of India to look for new partners for economic development. As passive reliance on the West would be detrimental to its national interests, Government of India came out with the Look East policy in the 1990s, expressing its desire to revive economic relations with countries in the East and Southeast Asian nations. Realizing the limitations of the South Asian Association for Regional Cooperation (SAARC), India took initiatives to start the Bay of Bengal Initiative for Multi-Sectoral Technical and Economic Cooperation (BIMSTEC), the Bangladesh–China–India–Myanmar (BCIM) Forum for Regional Cooperation and the Bangladesh–Bhutan–India–Nepal (BBIN)

Initiative. The Look East policy, although initiated by late Prime Minister P. V. Narasimha Rao in the 1990s, received a push during the United Progressive Alliance (UPA) regime in the first decade of the 21st century after Dr Manmohan Singh became the prime minister. The shift in policy brought the hitherto neglected NER into the centre of the development discourse (Sarma and Choudhury 2017). The policymakers who had earlier viewed the NER as a landlocked region realized the potentialities of the region to connect mainland India with the East and the Southeast Asian countries through taking advantage of the land link. The change in security and development priorities led to an emphasis on greater attention towards Northeast India. After the Narendra Modi-led National Democratic Alliance (NDA) came to power in 2014, the central government rechristened the policy as the Act East policy and laid stress on building border roads, especially in the state of Arunachal Pradesh bordering China, and strengthening India's relations with Association of Southeast Asian Nations (ASEAN) countries (Das and Thomas 2018). Apart from the efforts to boost trade between the north-eastern states and the countries bordering them, the governments have also taken initiatives to revive border *haats* to meet the needs of the people living in the borderlands. They have begun taking greater interest in addressing the socio-economic problems of the border people, keeping in view the significance of peace and development in the region.

NEED FOR MORE STUDIES OF BORDERLANDS OF THE NORTH EASTERN REGION

In the context of these changes and initiatives, it is necessary to aim at a comprehensive understanding of the political economy of the borderlands of India's NER, which would enable us to comprehend the historic links that the states and the communities in the region have with the neighbouring countries and assess their potentialities in boosting sub-regional cooperation between India and its neighbours. The Department of Political Science in North-Eastern Hill University, Shillong, as part of the University Grants Commission's (UGC) Special Assistance Programme (SAP), has been undertaking research and organizing national and international seminars to understand the potentialities and limitations of the region to become a link connecting India to its eastern neighbours. Recent years have witnessed the publication of several books and articles on India's Look East policy and Northeast India (Baruah 2003; Das and Thomas 2018; Muni and Mishra 2018; Sarma and Choudhury 2017). Land connectivity and economic development in and through the NER are the primary focus of these works. To understand the prospect of the Look East/Act East policy, one needs to examine the relations that India, Northeast India in particular, has with the neighbouring countries. Further, land connectivity with the neighbouring countries would make sense only if the people living in the borderlands live in peace and they become partners in sub-regional development. Hence, several issues of concern to the people in the north-eastern border-lands, such as poverty, underdevelopment, road connectivity, illegal migration, refugee problem, market accessibility, education and social welfare, become important. Keeping these aspects in mind, an effort is made in this volume to collate well-researched papers that throw light on different dimensions of the north-eastern borderlands and their relations with the neighbouring countries.

LOOK EAST/ACT EAST POLICY AND POLITICAL ECONOMY OF THE NORTH EASTERN REGION

The volume has two sections. The first section focuses primarily on the politics of economic development in Northeast India and the border-lands. The first two chapters paint a comprehensive picture of India's relations with its eastern neighbours, the political economy of the NER and border trade between north-eastern states and the countries bordering them. The chapters in the first section throw light on the potentialities of the region and the impact of government policies, set the agenda for revitalizing the region and explore the possibilities of building sub-regional economic cooperation with the countries neighbouring them. The second section focuses on inter- and intra-community interactions that take place in the borderlands. It throws light on variants of border crossings by the returnees, refugees and illegal migrants and the response of the local communities. The section looks at both conflicts and cooperation between the communities.

In the first chapter, 'Northeast India as Engine of India's Growth: Reorienting the Act East Policy Goals', Mahendra P. Lama probes why, despite the creation of so many legal and institutional mechanisms for the regional development, Northeast India lags behind the rest of the country. He expounds how such factors as the partition, physical disconnection, underdevelopment, neglect, insurgency, misgovernment and obsession with security concerns have left the NER underdeveloped and backward. The author opines that the projection of the NER as a region of conflict, violence, insurgency and instability discouraged the participation of the private sector, multi-sector organizations and non-state actors in playing a positive role in the development of the NER. According to him, federal priorities and the central government's obsession to bring the NER into the national mainstream have made the region dependent and backward, and he asserts that these syndromes and practices that made the NER a periphery should be corrected through what he calls 'reverse integration'. Although the policymakers are yet to heed to Mahendra Lama's ideas on making the NER the engine of India's growth story, in recent decades the government has acknowledged the need for developing the infrastructure in the NER. Rakhee Bhattacharya, in her chapter, 'Political Economy of Road Infrastructure in India's Northeast since Liberalization' examines the implications of India's structural shifts in its economic policy and their impact on the Northeast. The chapter interrogates the mega road projects in the NER against the background of India's security concerns and its economic compulsions to reach out to its neighbouring countries. While endorsing the need for re-imagining the NER as a gateway to Southeast Asia, the author expresses her apprehension that this neoliberal approach to the development of the Northeast would reify the centre–periphery relations in new forms and pleads for an inclusive and informed strategy for protecting the interests of the people and the region.

INDIA'S EASTERN NEIGHBOURS AND DEVELOPMENT OF THE NORTH EASTERN REGION

The first section also has chapters that provide an overview of India's relations with its eastern neighbours bordering the north-eastern states and examine their implications for the development of the NER.

The section emphasizes politics of economic development in the NER and in the neighbouring countries. Apart from the views of Indian scholars, the section also includes scholarly perspectives from Bangladesh, Bhutan, Nepal and China. The authors have made efforts to present national and regional perspectives, reflected on what should be done to promote sub-regional cooperation between India and its eastern neighbours and examined the implications of such cooperation for the states and the people in the NER.

Despite religious, cultural and historical links, the relations between India and Nepal have witnessed ups and downs. Some border areas such as Kalapani, Lipulekh, Susta are claimed by both countries (Kansakar 2002; Orton 2010). Sikkim is the only north-eastern state that shares a border with Nepal. Bishnu Dev Pant's chapter, 'Irritants in India–Nepal Relations and Potential Benefits of Greater Cooperation', throws light on the uneasy relations between India and Nepal. Reflecting on the irritants in the bilateral relations, the author calls for strengthening Nepal–India relations based on equality and mutual respect. The chapter also underscores the need for Nepal to engage directly with the north-eastern states of India. Unlike its relations with Nepal, India has maintained steady relations with Bhutan. Modifying the earlier 1949 Friendship Treaty that allowed India to guide Bhutan in foreign policy and defence matters, the two countries signed a new friendship treaty in 2007 which reiterates that both countries shall cooperate closely with each other on issues relating to their national interests, and that neither government shall allow the use of its territory for activities harmful to the national security and interests of the other (Nga et al. 2019). The open border that India has with Bhutan enables the citizens of either country to enter the neighbouring country without a visa. Bhutan borders the north-eastern states of Sikkim, Assam and Arunachal Pradesh. Some militant groups from the NER took advantage of the open border to operate their camps in the border areas of Bhutan. In 2003, the government of Bhutan cooperated with Government of India in carrying out an anti-insurgency operation against the Northeast militants hiding in the borderlands of Bhutan (Banerjee and Laishram 2004; Prabhakara 2014). To neutralize the militant activities, Government of India is building a border road along the India–Bhutan border and is contemplating erecting fencing in certain sensitive areas (Roy 2018). In her chapter, 'Indo-Bhutanese Relations: Evolution and Changing

Contexts', Yedzin W. Tobgay recounts the changing sociopolitical dynamics in Bhutan and throws light on the unique path that Bhutan has taken towards modernity and democracy. The chapter analyses the internal and external factors that compel Bhutan's dependency on India, despite the increasing voices within Bhutan for more honourable and balanced relations with India.

During the pre-colonial and colonial periods, different communities and territories in the NER had trade connectivity with the rest of India through East Bengal, now Bangladesh; four north-eastern states—Assam, Meghalaya, Tripura and Mizoram—border Bangladesh. However, the partition severed the NER's access to and through East Pakistan/Bangladesh. The NER has to use the Siliguri Corridor to get connected to Kolkata and beyond. Highlighting the travel cost and travel time involved in carrying out trade through the Siliguri Corridor, Abu Hena Reza Hasan and Sayada Jannatun Naim, in their article, 'Bangladesh and Northeast India: Planning Communication Links under GATT Article V', argue that India would benefit by making Bangladesh a land bridge for cost-effective implementation of India's Look East policy. As transit through Bangladesh would reduce time and cost, the chapter proposes mutually beneficial transit negotiation within the framework of Article V of the General Agreement on Tariffs and Trade (GATT).

There is considerable academic literature on the Sino-Indian border dispute and its implications for the NER, especially for Arunachal Pradesh. Although there were clashes between the Indian Army and China's People's Liberation Army (PLA) at Nathu La and Cho La in Sikkim (1967), Doklam (2017) and the Galwan Valley (July 2020), which are matters of concern, it should also be acknowledged that India has not fought any major war with China in the Northeast after 1962. Since the 1980s, both India and China have been engaged in negotiations to resolve the border dispute and to promote economic cooperation. For long, the landlocked nature of the NER has been one reason for its backwardness. China also has landlocked provinces that have remained peripheral and underdeveloped. In recent decades, the government of China has taken proactive measures to develop such peripheral provinces and create the necessary infrastructure for trade with neighbouring countries (Bhattacharya 2013; Guangsheng 2016). China's development strategy helped the landlocked Yunnan province emerge as a fulcrum for sub-regional development. Narrating the experience of Yunnan, Hu Xiaowen, in her chapter, 'Yunnan and Northeast India: Chinese Perspective on Sub-regional Development', shows how Yunnan's experience is relevant to the development of the territories covered under the BCIM Corridor. She argues that effective interaction and coordination between China's Yunnan province and the north-eastern states in India would benefit both countries.

Myanmar is another country that has an open border with India. India shares a 1,643-km-long open border with Myanmar. Although India and Burma cooperated with each other to give shape to the Non-Aligned Movement, subsequent to the military takeover of Burma, for different reasons, India did not evince interest in strengthening its relations with its neighbour, which rechristened itself as Myanmar, for decades. However, its pursuit of the Look East policy, now the Act East policy, compelled India to engage with Myanmar, as India understands the importance of Myanmar in reaching out to Southeast and East Asian countries. Munmun Majumdar's chapter, 'The Northeast and Myanmar: India–Myanmar Engagement in the Modi Years', focuses on the implications of the proactive role taken by Government of India under Prime Minister Narendra Modi in promoting political, economic and strategic engagement with Myanmar.

TRADITIONAL BORDER TRADE AND BORDER *HAATS*

Often, the national- and regional-development strategies overlook the problems and perspectives of the people living in the borderlands. The people in the borderlands of the NER suffer locational disadvantages, such as inaccessibility, unavailability of essential goods, absence of markets for local goods, etc. To buy or sell goods, the people in the borderlands have to travel long distances to the towns and cities in their respective states. It is easier for them to meet their basic needs by trading with the communities on the other side of the border. Prior to the drawing of national borders, traditional *haats* were operating in different parts of the Northeast, catering to the needs of the local populations

(Bhattacharjee 2000; Kakoti 2008). After independence, the securitization of the national borders virtually put an end to the operation of the traditional *haats*, much to the disadvantage of the people living along the borders. The border trade that takes place between countries through national highways and land ports, or land customs stations, has little relevance to the needs of the people living in the borderlands. Realizing the needs of the local population, efforts have been made in recent years to start border *haats* and promote border trade between borderland communities. The latter section of the book discusses the experience of border *haats* and border trade in Sikkim, Tripura and Meghalaya and evaluates the future prospects of such border *haats* in the borderlands of the NER.

Many colonial records show that different communities in the NER had brisk trade relations with neighbouring territories. The chapter 'Trade in Pre-colonial Arunachal Pradesh and Tibet', written by Amrendra Kumar Thakur, describes the trade relations of the tribes of Arunachal Pradesh with the people of the neighbouring areas in Tibet (now China). It provides a critique of the colonial conception of so-called primitivism and isolation of the tribes of Northeast India and examines the trade relations of the tribes of Arunachal Pradesh with the people of neighbouring areas in Tibet. Like the tribes of Arunachal Pradesh, the people of Sikkim also had close contacts with Tibet. Sikkim was a protectorate state of British India; the colonial authority used Sikkim as a conduit for trade with Tibet. Dechen Bhutia's chapter, 'Reviving Border Trade and Tourism along Nathu La in Sikkim', shows how Sino-Indian conflict in the post-war world compelled India to close Jelep La, halting the brisk trade taking place between Sikkim and Tibet in the pre-colonial and colonial eras. The move adversely affected the indigenous communities in Sikkim which depended on the cross-border trade and business activities. The author argues that the agreement to revive Nathu La, apart from normalizing the relations between India and China, benefits the border communities through promoting trade and tourism in the region.

During the colonial period, the then princely state of Tripura maintained cultural and economic relations with East Bengal. However, the partition of India and the influx of Bengali refugees into Tripura compelled India to securitize Tripura's border with East Pakistan, now Bangladesh. The international border obstructed the socio-economic interactions and affected the livelihood of the people living in the borderlands. The governments' decision to revive traditional border *haats* along the Tripura–Bangladesh border came as a relief to the border communities. Based on her field survey, Suparna Bhattacharjee, in her chapter titled 'Tripura–Bangladesh Borderlands: Socio-economic Significance of Border *Haats*', discusses the functioning of the two border *haats*—the Kamalasagar–Tarapur and Srinagar–Chagalnaiya *haats*—in Tripura along the India–Bangladesh border. The chapter also focuses on the impact of the border *haats* on the border communities and makes certain suggestions for improvement. Similarly, there was a flourishing trade between the hill communities of the NER and the plains of Bengal during the colonial period. Rakhal Kumar Purkayastha, in his chapter titled 'Meghalaya–Bangladesh Border Trade: A Study of the Balat–Sunamganj *Haat*', examines the trade relations that the Khasis, Jaintias and Garos had with the people of the plains of Sylhet during the colonial and pre-colonial eras and shows how the partition and the creation of international borders had adversely affected these tribal communities. Appreciating the government's efforts to open up border *haats* to promote trade between the border communities, the author shares his observations on the actual working of the Balat–Sunamganj *haat* in Meghalaya and suggests measures for addressing the problems encountered by different stakeholders involved in the border trade.

BORDER CROSSINGS: MIGRANTS, REFUGEES AND CRIMINALS

There is considerable literature on the refugees and illegal migrants from East Pakistan/Bangladesh, and on their impact on economy and politics in the north-eastern states like Assam and Tripura (Debbarma 2005; Hazarika 2000; Sinha 1998, 319–333). To avoid repetition, the volume consciously avoids discussion on the refugees and illegal migrants from Bangladesh and focuses on the border crossings that take place along the Indo-Myanmar and Indo-Nepal borders. The latter section includes chapters on border-crossers—migrants, refugees and criminals—in the borderlands. Tejimala Gurung Nag, in her chapter, 'Nepali/Gorkha Settlers in Northeast India: Colonial Encounters and Post-colonial

Dilemmas', addresses the question of how Nepali-speaking people have become Indian citizens. She explains how Nepal's encounter with British India facilitated the migration and settlement of Gorkha soldiers and Nepali herders in different states in Northeast India.

Apart from the legal and illegal migrants, the NER was compelled to accommodate refugees (Ghoshal 2020; Singh 2009) from Bangladesh, Tibet and Myanmar. There are quite a few studies on the refugees from East Pakistan/Bangladesh, but not much attention is given to the Indian communities and Indian refugees from Myanmar. When Burma became part of British India, the colonial rulers encouraged several Indian communities to migrate to and settle in Burma. The Indian migrants played a very important role in the development of Burmese economy. However, with the rise of Burmese nationalism, the native Burmese's perception of Indian communities as their competitors and enemies led to the marginalization and even deportation of Burmese Indians. Emdorini Thangkhiew's chapter, 'Burmese Indians: Growth of Burmese Nationalism and Ethnic Discrimination', examines the existential dilemmas of the Burmese Indians and raises questions over the future of the Indian community still living in Myanmar. It also attempts to examine Government of India's policy towards the Indian diaspora in Burma. The chapter by Saurabh Kaushik, titled 'Returnees and Refugees from Burma to India: Differing State Responses', highlights the patterns and trends of refugee inflow from Burma (Myanmar) to India over the years. The chapter examines the trajectories of Indian-origin returnees (mostly Tamils), the Chin and the Rohingya. It brings to light the differing state responses to these three distinct groups of refugees. Underscoring the differential treatment, the author makes a plea for institutionalization of a coherent refugee regime in India based on international law.

Apart from refugees and illegal migrants, one can see ethnic militants, traffickers and smugglers of different types crossing the international borders. There are several studies on the parallel economy operating across the borders involving the trafficking of women, drug trafficking, smuggling of gold, arms and cattle that take place across India's borders with Bangladesh, Myanmar and Nepal (Bhaumik 2009; Dasgupta 2001; Nag 2002b; Nepram 2002; Sahni 2001). Such criminal activities become justification for the states to securitize their borders. H. Srikanth and T. T. Haokip's joint chapter 'Drug Trafficking in and through Myanmar and Manipur' explores the nature and magnitude of the problem of drug trafficking across the Manipur–Myanmar border and examines how the governments justify their decision to fence the international border by citing the effects of drug trafficking on the lives of the people in Manipur.

INTERACTIONS AMONG INDIGENOUS PEOPLES IN THE BORDERLANDS

The separation of Burma from British India and the partition of British India into India and Pakistan led to the redrawing of territorial boundaries in the subcontinent, creating artificial barriers separating ethnic communities speaking a similar language and practising a similar culture. Apart from the Bengali community, several indigenous ethnic communities, such as the Nagas, Kuki-Chin, Khasis, Garos, etc., also found themselves on both sides of the international borders. Despite their cultural similarities, these indigenous communities are forced to live under different national jurisdictions that impose restrictions on their movement through securitizing the borders. It is interesting to study the interactions, dilemmas and changing identities of such indigenous communities living in the borderlands.

The volume presents three chapters that discuss the experiences of three indigenous communities separated by the international border. In their chapter 'Understanding the Kukis' Opposition to Fencing of the Indo-Myanmar Border', T. T. Haokip and H. Srikanth call attention to the interactions and interdependence among the Kukis living in the borderlands of Moreh (India) and Tamu (Myanmar). The chapter presents the perceptions of the Kuki communities in Moreh and Tamu on the governments' decision to fence the Manipur–Myanmar border. The field report presented in the chapter shows commonalities that continue to bind the Kukis on either side of the international border. In contrast, Jelle J. P. Wouters' chapter, titled 'Borders That Divide: Naga Ethnoscape and Idea of Supranational Citizenship', shows that despite ethnic and cultural similarities, once the national borders are drawn, ethnic communities tend to develop distinct regional identities, which may not be compatible with ethnic solidarity. Wouters observes that after the partition, the Garos acquired distinct identities as Indian Garos and Bangladeshi Garos. However, in the case of the Nagas, there is still a strong demand for integration of all Naga-inhabited areas of India and Myanmar, despite constant bickering among the Naga tribes divided along regional lines. Wouters proposes the idea of supranational citizenship that accommodates the aspiration of the Nagas for cultural unity without disturbing the national political and territorial divisions. Continuing the discussion on the impact of national borders on the native communities, R. K. Satapathy and P. C. Lalthansiami, in their chapter, 'Chin Migration to Mizoram: Ethnic Affinity and Changing Perceptions', discuss the changing relations between the Mizos of Mizoram and the Chin of Myanmar, both belonging to the same ethnic stock. The chapter explores the factors that have led to the trans-border migration of Chin from Myanmar to Mizoram and analyses the reasons for the growing resentment among the Mizos against the Chin migrants.

SIGNIFICANCE OF THE PROJECT

For decades, India has focused more on the security threats emanating from Pakistan and China on its western and northern fronts. However, internal and external compulsions forced India to turn its attention towards its eastern neighbours. India's eastward strategy brought into focus the importance of its NER. The NER borders five countries. All north-eastern states of India have at least one country bordering them. The borderlands that each of the north-eastern states shares with a country or countries have a unique history, ethnic composition and social dynamics. Some native tribes that are now categorized as borderland communities had cultural and trade relations with the people in the neighbouring territories. Near absence of restrictions on the movement of the peoples enabled cultural and trade interactions in pre-colonial times. After the British established their control over the region, in their own colonial interests, the colonial rulers laid roads connecting the NER with Bengal and the neighbouring countries like Sikkim, Tibet and Nepal. Some communities and territories were no doubt left untouched, but the then-composite province of Assam witnessed considerable economic development during the period. Even the princely states of Manipur and Tripura had cultural and economic contacts with their neighbouring territories.

The situation, however, changed in the 20th century with the bifurcation of British India, the birth of nation states and the demarcation of national boundaries. These developments turned the NER into a landlocked and peripheral region. The Sino-Indian War of 1962, the rise of ethnic insurgencies and the flow of refugees and illegal migrants into the region made policymakers look at the region mainly from a security perspective. The funds pumped for economic development of the region benefitted the urban dwellers and contributed little to the economic and infrastructural development in the peripheral areas. The concentration of power in the hands of the central government and the securitization of the borderlands curtailed the autonomy of the states to address their problems and explore alternative strategies for development. Given the decades of stagnation and neglect, many native people in the NER welcomed India's Look East policy, now rechristened as the Act East policy. There is expectation that the new orientation would put an end to the artificial restrictions imposed by the nation states and help in spurring growth and development in the landlocked NER. The people's hopes and expectations found space in multiple books and articles published on the Look East/Act East policy.

The present volume takes the discussion further, focusing on the issues that are overlooked. It seeks to provide a comprehensive understanding of India's relations with its eastern neighbours and examine the place of the north-eastern borderlands. Apart from discussing the economic issues, the volume emphasizes historical, political and sociocultural issues. Considerable attention is given to the issues of border communities and borderland issues. The scholars who have contributed to the volume come from different disciplines—History, Political Science, Economics, International Relations, Anthropology and Refugee Studies. The inclusion of chapters from scholars from neighbouring countries helps in understanding their perspectives on sub-regional politics and development initiatives. While some chapters are empirical and descriptive, there are others which are analytical and critical. The authors come from different regional and ideological backgrounds. The editors did not direct the discussion to one line of thought. Still, most contributors to the edited volume, while welcoming the governments' initiatives to strengthen sub-regional cooperation with the eastern neighbours, emphasize reorienting the goals of India's Act East policy keeping in view the interests and concerns of the people in the borderlands of India's Northeast.

Whether the governments heed the scholarly perspectives and suggestions depends on many factors. India's emphasis on infrastructural development in the NER is motivated partly by security concerns and partly by its neoliberal strategy to reach out to Southeast Asia and the Indo–Pacific region. The attempts to explore and expand markets for the goods produced by the monopolies may no doubt make the NER a transit corridor, but they may not automatically contribute to the development of the region. The NER's success depends on how the communities and the governments in the region articulate their concerns and reorient the policy goals to serve the interests of the region and the border communities. Further, the sub-regional development in the region is also contingent on how far India overcomes its preoccupation with the security concerns. The BCIM initiative may not yield the intended results, as India is unlikely to overcome its fear of Chinese domination. As far as other projects, such as BBIN and BIMSTEC, are concerned, their success depends on how far the countries in the region can overcome the global recession. In the absence of other conducive factors, road connectivity alone may accelerate economic growth. The influx of illegal migrants and refugees has made the native people apprehensive of any developmental activity likely to attract outsiders into the region. Although ethnic insurgencies appear to be under check at the moment, the perpetuation of xenophobia stands in the path of the governments' plans to initiate major industrial and developmental projects in the region. It is therefore necessary to initiate appropriate development strategies, such as trade through border *haats* and community tourism, that benefit the locals and, at the same time, not rouse xenophobia.

The volume has made efforts to capture some of these essential dynamics and contradictions of India's relations with its eastern neighbours and their likely impact on the borderlands of the NER. However,

India's north-eastern borderlands are so varied and complex that it is not possible to encapsulate all dimensions of the borderlands in one volume. There is a need for many more border studies on local economies, border policing, citizenship, inter-ethnic relations, migration, refugee influx, gender, environment and the impact of epidemics on the borderlands of the NER. We view the volume as one step towards a comprehensive understanding of the dynamics of the borderlands of the NER.

▸ REFERENCES

Andreas, P. 2002. 'Transnational Crime and Economic Globalization'. In *Transnational Organized Crime and International Security: Business as Usual*, edited by M. Berdal and M Serrano, 37–55. Boulder: Lynne Rienner Publishers.

Arpi, Claude, 2013. *1962 and the McMahon Line Saga*. New Delhi: Lancers.

Banerjee, D., and B. S. Laishram. 2004. *Bhutan's 'Operation All Clear': Implications for Insurgency and Security Concern* (IPCS Issue Brief 18). 1–4 January. http://www.ipcs.org/pdf_file/issue/IB18-OperationAllClear.pdf (accessed 30 September 2020).

Baruah, Sanjib. 2003. *Between South and South East Asia: Northeast India and the Look East Policy* (CENISEAS, Paper No. 4). Guwahati: OKDISCD.

Baruah, Sanjib. 2020. *In the Name of the Nation: India and Its Northeast*. Stanford: Stanford University Press.

Bhattacharjee, J. B. 1990. 'The Sylhet Referendum (1947): Myth of a Communal Voting'. *Proceedings of the Indian History Congress* 51:482–487.

Bhattacharjee, J. B. 2000. *Trade and Colony: The British Colonization of North-East India*. Shillong: NEIHA.

Bhattacharjee, Nabanipa. 2009. 'Unburdening Partition: The "Arrival" of Sylhet'. *Economic and Political Weekly* 44(4):77–79.

Bhattacharya, Abanti. 2013. *China and Its Peripheries: Strategic Significance of Tibet* (Issue Brief 220). 10 May. New Delhi: Institute of Peace and Conflict Studies.

Bhaumik, Subir. 2009. *Troubled Periphery: The Crisis of India's North East*. New Delhi: SAGE Publications.

Ceglowski, Janet. 1998. 'Has Globalization Created a Borderless World?' *Business Review*, March/April, 17–27.

Cox, Lloyd. 2004. 'Border Lines: Globalisation, De-Territorialisation and the Reconfiguring of National Boundaries'. Sydney: Centre for Research on Social Inclusion, Macquarie University.

Das, Gurudas, and C. Joshua Thomas, eds. 2018. *Look East to Act East Policy: Implications for India's Northeast*. New York: Routledge.

Das, Pushpita, ed. 2010. *India's Border Management: Select Documents*. New Delhi: IDSA.

Dasgupta, Anindita. 2001. 'Small Arms Small Arms Proliferation in India's North-East: A Case Study of Assam'. *Economic and Political Weekly* 36(1):59–65.

de Wenden, C. W. 2020. 'Borders and Migrations: The Fundamental Contradictions'. In *Migration, Borders and Citizenship: Between Policy and Public Spheres*, edited by Maurizio Ambrosini, Manlio Cinalli, and David Jacobson, 47–60. London: Palgrave.

Debbarma, K. 2005. 'Inter-Ethnic Conflict in Tripura: Causes and dimensions'. In *Inter-Ethnic Conflict in Northeast India*, edited by Girin Phukon, 141–147. New Delhi: South Asian Publishers.

Diener, Alexander C., and Joshua Hagen. 2012. *Borders: A Very Short Introduction*. Oxford: Oxford University Press.

Dutta, Binayak, and Suranjana Choudhury. 2017. *The forgotten partitions of northeast India and its lingering legacies*. 15 August. New York: Café Dissensus.

Farrelly, Nicholas. 2012. 'Review of the Noboru Ishikawa's book, *Between Frontiers: Nation and Identity in a Southeast Asian Borderland*'. *South East Asia Research* 20(4):628–630.

Ghoshal, Anindita. 2010–2011. 'Survival Question of East Bengal Refugees: The Case of Tripura (1946–71)'. *Proceedings of the Indian History Congress* 71:1208–1215.

Ghoshal, Anindita. 2020. *Refugees, Borders and Identities: Rights and Habitats in East and Northeast India*. New Delhi: Routledge India.

Gogoi, Dilip. 2020. *Making of India's Northeast: Geopolitics of Borderland and Transnational Interactions*. New York: Routledge.

Guangsheng, Lu. 2016. *China Seeks to Improve Mekong Sub-Regional Cooperation: Causes and Policies*. S. Rajaratnam School of International Studies. 1 February.

Hämäläinen, Pekka, and Samuel Truett. 2011. 'On Borderlands'. *The Journal of American History* 98(2):338–361.

Hazarika, Sanjoy. 1994. *Strangers of the Mist: Tales of War and Peace from India's Northeast*. New Delhi: Penguin India.

Hazarika, Sanjoy. 2000. *Rites of Passage: Border Crossings, Imagined Homelands, India's East and Bangladesh*. New Delhi: Penguin India.

Kakoti, Sanjeeb. 2008. 'Trade and Trade Routes: Historical Perspectives'. In *Indo-Bangladesh Border Trade: Benefitting from Neighbourhood*, edited by Gurudas Das and C. J. Thomas, 312–331. New Delhi: Akansha Publishing House.

Kansakar, Vidya Bir Singh. 2002. *Nepal–India Open Border: Prospects, Problems and Challenges*. Kathmandu: *Institute of Foreign Affairs*.

Kent, George. 2017. 'When Burma and India went their separate ways'. *Myanmar Frontier*. 5 September. https://www.frontiermyanmar.net/en/when-burma-and-india-went-their-separate-ways/ (accessed 30 September 2020).

Kitching, Gavin. 2001. 'The role of transnational corporations'. In *Seeking Social Justice through Globalization: Escaping a Nationalist Perspective*, edited by Gavin Kitching, 33–48. Pennsylvania, PA: Penn State University Press.

Leake, Elisabeth. 2016. *The Defiant Border: The Afghan-Pakistan Borderlands in the Era of Decolonization, 1936–1965*. Cambridge: Cambridge University Press.

Lessman, Thomas. 2015. 'The Changing Map of India from 1 AD to the 20th Century'. *Scroll*. 3 May. https://scroll.in/article/722369/the-changing-map-of-india-from-1-ad-to-the-20th-century (accessed 30 September 2020).

Marx, Karl. 1853. 'The British rule in India'. *New York Daily Tribune*, 6 June. https://www.marxists.org/archive/marx/works/1853/06/25.htm (accessed 30 September 2020).

Maxwell, Neville, ed. 2015. *India's China War*. New Delhi: Natraj Publishers.

Ministry of Home Affairs. 2019. *Annual Report 2018-19*.

Misra, Sanghamitra. 2001. *Becoming a Borderland: The Politics of space and Identity in Colonial Northeastern India*. New Delhi: Routledge.

Mukherjee, Kunal. 2019. *Conflict in India and China's Contested Borderlands: A Comparative Study*. New York: Routledge.

Muni, S. D., and Rahul Mishra. 2018. *India's Eastward Engagement: From Antiquity to Act East Policy*. New Delhi: SAGE Publications.

Nag, Sajal. 2002a. *Contesting Marginality: Ethnicity, Insurgency and Sub-Nationalism in North-East India*. New Delhi: Manohar.

Nag, Sajal. 2002b. 'Structure of a Non-state Economy: Political Economy of North-east India'. *Contemporary India* 1(4).

Nayak, Sohini. 2020. 'India and Nepal's Kalapani border dispute: An explainer'. *ORF Issue Brief*, April 29. https://www.orfonline.org/research/india-and-nepals-kalapani-border-dispute-an-explainer-65354/ (accessed 30 September 2020).

Nepram, Binalakshmi. 2002. *South Asia's Fractured Frontier: Armed Conflict, Narcotics & Small Arms Proliferation in India's Northeast*. New Delhi: Mittal Publications.

Nga Le Thi Hang, Tran Xuan Hiep, Dang Thu Thuy, and Ha Le Huyen. 2019. 'Bhutan Treaties of 1949 and 2007: A Retrospect'. *India Quarterly* 75(4):441–455.

Orton, Anna. 2010. *India's Borderland Disputes: China, Pakistan, Bangladesh, and Nepal*. New Delhi: Epitome Books.

Parker, Noel, and Nick Vaughan-Williams. 2014. *Critical Border Studies: Broadening and Deepening the 'Lines in the Sand' Agenda*. London: Routledge.

Prabhakara, M. S. 2014. 'Crackdown in Bhutan'. *Frontline* 21(1):3–16.

Ramunny, Murkot. 1988. 'Changing Face of Tripura'. *Economic and Political Weekly* 23 (37):1879–1880.

Ranjan, Amit. 2018. *India–Bangladesh Border Disputes: History and Post-LBA Dynamics*. New York: South Asia Economic and Policy Studies, Springer.

Roy, Anirban. 2018. 'Delhi mulls constructing border road, barbed fence along 699-km Bhutan border'. *The Economic Times*, 1 August.

Rytövuori-Apunen, Helena, and Renée Marlin-Bennett. 2016. 'Epilogue: Charting Border Studies Beyond North American Grounds'. In *The Regional Security Puzzle around Afghanistan: Bordering Practices in Central Asia and Beyond*, edited by Helena Rytövuori-Apunen, 285–296. Berlin: Barbara Budrich Publishers.

Saddiki, Said. 2017. *World of Walls: The Structure, Roles and Effectiveness of Separation Barriers*. Open Book Publishers.

Sahni, Ajai. 2001. 'The Terrorist Economy in India's Northeast: Preliminary Explorations'. *Faultlines* 8. https://www.satp.org/satporgtp/publication/faultlines/volume8/Article5.htm (accessed 30 September 2020).

Saikia, Pahi, and Anasua Basu Ray Chaudhury. 2019. *India and Myanmar Borderlands: Ethnicity, Security and Connectivity*. New Delhi: Routledge India.

Sarma, Atul, and Saswati Choudhury. 2017. *Mainstreaming the Northeast in India's Look and Act East Policy*. New York: Springer.

Sethi, Sunil. 1978. 'Did India have a right to annex Sikkim in 1975'. *India Today*, 30 April. https://www.indiatoday.in/magazine/cover-story/story/19780430-did-india-have-a-right-to-annex-sikkim-in-1975-818651-2015-02-18 (accessed 7 Feb 2021).

Sidhu, G. B. S. 2018. *Sikkim—Dawn of Democracy: The Truth behind the Merger with India*. New Delhi: Penguin Viking.

Singh, Deepak K. 2009. *Stateless in South Asia: The Chakmas between Bangladesh and India*. New Delhi: SAGE Publication.

Sinha, S. K. (1998). Report on Illegal Migration to Assam submitted to President of India, November 8, 1998'. In *India's Border Management: Select Documents*, edited by Pushpita Das, 319–353. New Delhi: ICPS.

Upadhyay, Archana. 2019. *India's Fragile Borderlands: The Dynamics of Terrorism in North East India*. New Delhi: Routledge.

van Schendel, Willem. 2005. *The Bengal Borderland: Beyond State and Nation in South Asia*. London: Anthem Press.

Wastl-Walter, Doris. 2019. *The Ashgate Research Companion to Border Studies*. Farnham: Ashgate.

William, P., and D. Vlassis. 2005. *Combating Transnational Crime: Concepts, Activities and Responses*. London: Frank Cass.

Wilson, Thomas M., and Hastings Donnan. 2012. *A Companion to Border Studies*. New Jersey: Wiley Blackwell.

Wonders, Nancy A. 2007. 'Globalization, Border Reconstruction Projects, and Transnational Crime'. *Social Justice* 34(2), 33–46.

PART I

India's Relations with Neighbours: Politics of Development in NER

Chapter 1

Northeast India as Engine of India's Growth
Reorienting the Act East Policy Goals

Mahendra P. Lama

INTRODUCTION

The eight states of India's North Eastern Region (NER) have been bestowed with special institutional, development and financial provisions in the Constitution, which assigns them a significant degree of federal autonomy. Under the special category states[1] introduced in 1969, the NER received liberal funding for both central sector and externally aided projects. Since 1996, the region has received at least 10 per cent of the plan budget(s) of the central ministries/departments under the new provision of Non-lapsable Central Pool of Resources (NLCPR). Some innovative interventions such as North East Industrial and Investment Promotion Policy (NEIIPP), which was renamed as North East Industrial Development Scheme in 2018, have been implemented since 1996.

Many specialized agencies such as the North Eastern Development Finance Corporation Ltd (NEDFi), North Eastern Regional Agricultural Marketing Corporation Ltd (NERAMAC), North Eastern Electric Power Corporation Limited (NEEPCO) and North Eastern

Handicrafts and Handlooms Development Corporation (NEHHDC) have been set up in the Northeast. International institutions, such as Japan International Cooperation Agency[2] (JICA) and International Fund for Agricultural Development[3] (IFAD) and multilateral institutions, such as the World Bank[4] and the Asian Development Bank (ADB),[5] are now major players in the development process. Government of India (2004) set up the North Eastern Council in 1971, which was restructured in 2004 to make it a regional planning agency. The Department of Development of North Eastern Region was established at the national level in September 2001 and was upgraded to a full-fledged Ministry of Development of North Eastern Region (MDoNER) in May 2004. This is the only ministry that has an exclusive territorial jurisdiction in India. Both the union government and state governments have come out with a range of reports that deal with various critical issues that beset the NER.[6] The NER Vision Document 2020, prepared and presented in 2008, was to a certain extent operationalized at both the national and state levels. The NER Vision Document 2035 is now being prepared by the National Institution for Transforming India (NITI) Aayog.

With all these provisions, interventions and federal devolution, one would have expected the NER to be a 'growth pole' and the pivot in the Indian economic scenario. The 'take off' should have taken place at least two decades back, but that did not happen. The region started well, as evident from the fact that Assam was one of the fastest growing states in pre-partition India, with much greater industrialization triggered by huge infrastructural interventions, like the railway network. The gross public investment in economic overheads (such as irrigation, power, roads and railways) amounted to about ₹17,187 million during 1860–1947. The NER absorbed the largest proportion, amounting to ₹5,207 million, or 30.3 per cent of the gross public investment in economic overheads triggered by investment in railways, which accounted for about 65 per cent of the total investment. The East Indian Railway was the largest single claimant (Thavaraj, 1972). The expanded connectivity was necessitated by the growth of tea plantations and harnessing of the mineral resources like oil and gas. Tea plantations in Assam increased from occupying a mere 1,880 acres in 1850 to occupying 20,460 acres in 1901, and tea production jumped from

2.16 lakh pounds to 72.38 lakh pounds, respectively, and exports rose from 2.89 lakh maunds in 1880 to 12.39 lakh maunds in 1900 (Nath, 2005). A World Bank strategy report stated, 'Steamships moved along the bustling Brahmaputra and Barak waterways to Calcutta carrying Assam tea to London auctions. Coal was mined and Digboi still boasts of the oldest producing oil well and refinery in the world' (World Bank 2007, 23).

UNDERSTANDING THE CAUSES OF UNDERDEVELOPMENT

Why, despite all these special constitutional, institutional and financial provisions and policy decisions, did the NER remain a relative laggard? There are scores of reasons for the sordid and shallow post-independence development trajectory in the NER. Such factors as the abrupt physical disconnection in the post-partition period[7] triggered by the separation of East Pakistan (now Bangladesh), insurgency- and violence-led instability, acute misgovernance, lack of vision and national and global outlook among the local leaderships and serious institutional laggardness have protracted this backwardness. The bullock cart–paced development of physical infrastructures and the widespread and endless identity politics added to this laggardness. The actors at the centre increasingly realized the failure of national policymakers to visit the region and the inability to understand and accept the potentialities of the communities in the NER. It was only recently that the National Democratic Alliance (NDA) government led by Prime Minister Narendra Modi issued a notification (dated 16 January 2015) asking the union ministers to tour the NER every fortnight.[8]

Besides the reasons mentioned in the above paragraph, the decision-makers and implementing agencies in New Delhi view national security as the core concern for undertaking developmental activities in the region. Since independence, all major development initiatives in the region had been overwhelmingly implemented and supervised by the national security–related agencies and ministries. The post-colonial official justification of these designs, approaches and strategies of development interventions with intrusive security parameters allowed the continuation of the time-tested colonial legacy (Bhaumik 2009; Pemberton 2015), confining the NER in the boxes such as 'excluded area', 'partially excluded area', 'inner line permit' and 'restricted area provisions' and to exclusivity. In the immediate aftermath of independence, there were highly biased apprehensions about the nationalistic traits and misplaced perceptions about the very political orientation of the people of the NER, as is evident from the letter Sardar Vallabhbhai Patel, the then home minister of India, wrote to Pandit Jawaharlal Nehru, the then prime minister of India, on 7 November 1950.[9] This genre of thinking remained protracted for a few decades to come.

A critical mass of political leadership, bureaucracy and even media in India nurtured the security-centric acculturation, perpetuating the idea that every experimentation and innovative development project in the NER should be first tested in the laboratories of security institutions. All this contributed to the evolution of the national discourse that presented the NER as a region beset with violent conflicts, insurgencies and instability and hence a 'no go' geography for the rest of India and also for others from foreign lands. Conflict-driven instabilities have been a reality in the NER. Once, the entire eastern frontier had been a zone of conflict and instability. If one stood in a place like Guwahati, within a 700-km perimeter, one could find all varieties of conflicts. No single and compact geographical region, possibly anywhere in the world, has recorded such a variety of conflicts, wars, violence and instabilities as has been recorded by the NER. Except nuclear war, one can name any genre of conflicts in the NER.[10]

Many of the insurgencies, movements and conflicts have had proven patronization from trans-border agencies (Weiner 1993). However, these conflicts have been highlighted and projected in a manner as if the entire NER region protractedly remained engulfed in fire and violence. However, the fact remains that only a miniscule and a negligible minority (perhaps less than 1 per cent of the total population) have been within the precincts of conflicts, insurgency and violence in this otherwise peaceful, community-led abode of harmony and tranquillity. Nevertheless, the national and global discourse on the NER ignored and marginalized 99 per cent of the people and communities who have been outside this conflicting tug of war. It only highlighted the miniscule elements that were responsible for the conflicts. In this case, some key actors and institutions gained, while the nation and the people at large lost a lot, including two–three younger generations. In the actual sense, the right policy practice should have been to make this 'over 99 per cent' the real constituency and the focal point for peace, stability and development. The opportunity was grossly overlooked.

Security has continued to elude the people. A recent NITI Aayog report mentioned that

> perception of people in the border areas regarding security issues varied across States. 50% people of Manipur, 82% people of Tripura and 14% people of Nagaland settled in these areas said they do not feel secure. Similarly, 78% people of Sikkim and 65% people of West Bengal said they do not feel secure living in border areas. (NITI Aayog 2015, 25–26)

Illegal foreign immigrants constitute another issue of concern in the region. If the staggering number of illegal foreign immigrants found by the National Register of Citizens (NRC) in Assam is to be believed, it also seriously calls into question the working of state- and national-level institutions. The most fundamental question is how and when these staggering numbers entered Assam and settled down over decades with no punitive action taken against them—which institutions are responsible, which officials are accountable and, more seriously, which are the exact border crossing points that have been infringed upon and violated. It is a reflection of comprehensive failure, including constitutional, institutional, official, security, legal–administrative and governance. All these years, the failure of state machineries and institutions to check illegal migration was literally swept under the carpet. And now, some institutions and machineries, who themselves were responsible for preventing and regulating movement of illegal foreign immigrants at the border points, are again involved in the tedious, costly and near-impossible process of detecting, determining and deporting the illegal immigrants. This lapse brings forth the importance of border management and emphasizes the need to understand what exactly attracts international migrants to cross over to Assam and the NER. It also brings forth the essentiality of legislating comprehensive in-migration laws and policies, including those related to refugees who cross the international border, which actually never happened in post-independence India.

DEBILITATING ISSUES

In the post-independence period of last seven decades, five core debilitating issues emerged, which resulted in severe impairment of human security, sharp erosion in governance practices, a newer variety of development disconnections and leakages, dislocation of political objectivity, shrinking of democratic space, displacement of the larger social–communal ethos and literal fragmentation of the rich culture of cross-border exchanges.

First, the national discourse about the NER became hegemonic, where one was made to believe that the NER is a region of conflict, violence, insurgency and instability—a region of rugged terrains and low self-sustaining capacity and ultimately a development laggard. Security issues overwhelmed the development, social, environmental and cultural dynamics of the region. This kept all the potential development partners, including the private sector, global multilateral agencies and non-state actors, away from any meaningful participation in the development process.

Second, the building of pivots and critical educational, professional, development and political institutions was de-prioritized and made to play second fiddle. In the absence of the institutions that connected the communities with the rest of India and the global world and which trained, built capacities and empowered the huge mass of talented youths in the region, a major opportunity was lost to galvanize the people's participation in the national development and integration process. Money was poured incessantly, but no investment was made on social capital formation and reconnecting of the region with the vast, newly industrializing countries in Southeast Asia and East Asia. While the neighbouring Southeast Asian countries were declared the 'flying geese', the NER in India still unwillingly celebrated its backwardness and alienation. The aspirations of youths got derailed and impregnated with the feeling of powerlessness.

Third, the NER as a geography and community began to become alienated from the crucial and broader matrices of the 'national mainstream'. Communities became inward-looking, identity politics became much more diabolical, and newer forms of protests and resistance emerged. Some of the latter even got prolonged support, training and habitat facilities in alien neighbouring countries that were against India's national sovereignty interests. On the other hand, the 'Us vs. Them' theme steadily mesmerized the psyche of dominant players in the 'centre' and satiated political leaderships in the 'periphery'. Despite the creation of several states and their own governance machineries, the level of confidence at the federal level on the ability and reach of the locals to manage the political economy of the region remained relatively low. This was blatantly witnessed in the parallel running of several security measures, policies and institutions. This crisis of confidence and trust deficit has been the biggest bottleneck in the forward development leap of the NER. Barring a few initiatives by astute politicians, talented bureaucrats and sagacious civil society members, no collective and convincing initiatives could be taken up by the actors and institutions in the NER.

Fourth, the borderlands which remained witness to large-scale interactions in the pre-independence period gradually became a 'no go' zone. This reduced and shrunk vibrant borderlands into just borders, more as a geometric line. And finally, the political leadership in most of the NER states lost the larger vision of the national and global horizon, and the democratic institutions started losing their manoeuvrability. While the nation celebrated the second-generation reforms, the NER remained only in the periphery of just the first-generation reforms. Institutions and actors in governance got entangled in the dilemma as to what they should emphasize—development dynamics or the security compulsions. Moreover, they became solidly, acutely dependent on federal funding and political dictates therein. The NER's 'regionness'—what Hettne (Bjorn 1999) described as a wholesome geographical unit, a social system, an organized cooperation, a civil society and an acting subject with a distinct identity—gradually got eroded. The entire development discourse became government-centric. This made the citizens too dependent on the state. This definitely had a visibly adverse impact on the competitiveness, creativity, innovation and outward-looking attitude in the society and economy, which was more serious among the youth population. This story and psyche of marginalization of the periphery remained and ruled for a full five–six decades after independence.

ACTING NORTHEAST

All these syndromes and practices that made the NER a periphery have to be corrected now. Act East (Look East of the early 1990s), initiated by Prime Minister Modi, has to be essentially based on acting Northeast. This is partly because the India–Southeast Asia integration very much intersects with local integration in the NER. Acting Northeast would have four-way effects. First, it would reactivate and reorient the development process within the NER and strengthen crucial interconnections among the NER states. Second, it would help connect the rest of India with the NER and more vigorously open up and make efficient use of transit facilities to the NER provided by Bangladesh. Third, it would help in realizing the larger goals of the Act East policy of India through prompting physical, virtual, commercial and sociocultural connectivity with the countries of Southeast Asia and East Asia. Thus, the Northeast would become the real connecting cultural ecology and not just remain a symbolic flyover. Fourth, it would attract Bangladesh, Nepal, Bhutan and Myanmar to join this eastward sojourn further, fostering the Bangladesh–Bhutan–India–Nepal Initiative (BBIN), Bangladesh–China–India–Myanmar Economic Corridor (BCIM) and Bay of Bengal Initiative for Multi-Sectoral Technical and Economic Cooperation (BIMSTEC) initiatives (Centre for Studies in International Relations and Development 2009; Rahman et al. 2007).

The orthodox model of expecting, enticing and forcing the NER to join the rest of India has largely failed. The reverse model has to be applied and practised now. China practised this reverse integration across the entire sub-regional landscape of its south-western region, including Sichuan, Tibet, Yunnan, Xinjiang and Qinghai. The nation state went to these relatively less developed frontiers after it developed the coastal best in the south-eastern region including through its noted 'develop the west' policy launched in 2000. What needs to be emphasized now in India is how the centre would integrate with the Northeast's periphery rather than the erstwhile model of merging the periphery with the core. The strategic community that continues to harp on the 'chicken neck' thought process and juxtapose the NER as a relatively small economy with 7.9 per cent of India's total geographical area, 3.8 per cent of the population and 4.02 per cent of the GDP has to undergo a visible metamorphosis. We must revisit the political geography. In the changing national and international contexts, it makes sense to view the NER as the chicken's head and the rest of India as its body. An awkward situation with inbuilt instability and inhibitions arises when the body guides and leads the head. This is what actually happened in the past. The best way forward is therefore to let the head lead the body so that we come back to nature. Such reverse integration unfurls development, creativity, innovations and inclusiveness. The NER is displaying its prowess to globalize the locals through at least sports[11] now. Mary Kom, Hima Das and Sunil Chhetri have emerged as new global brand ambassadors from the NER. Similarly, the traditional medicinal systems, genetic affluence and biodiversity, cross-cultural practices and also music, cinema and tourism corridors and climate change adaptations knowledge base in the region could be steadily branded and globalized.

Such reverse integration happened in Darjeeling during both the pre- and immediate post-independence period, globalizing the local tea, cinchona, a toy train and schools and colleges and attracting millions of tourists, experts and professionals from the rest of India and the world to Darjeeling. This tiny, strategically located district emerged as the core of the Tea Horse Road to Tibet and Yunnan in China through Jelep La and Nathu La, produced many Olympic players and a host of valiant fighters in the defence forces, hosted several Bollywood films and actors, earned millions of hard currencies from tea, quinine and timber and educated top-notch civil servants and other professionals in its schools and colleges, from both India and neighbouring countries. With its branding as the 'queen of the hills', Darjeeling became irreversibly globalized. Here, the rest of India and the global world first came to Darjeeling, and the latter steadily got integrated with their systems (Lama 1989; Sarkar and Lama 1986). The NER has to learn from the Darjeeling experience.

To spur development, the rest of India entering the NER should induct a variety of physical infrastructures, modern institutions, technology, sustainable investments, technical know-how and trans-border connectivity and sub-regional development projects, like energy corridor, war and Buddhist circuit tourism and value chains in organic products, traditional medicinal systems, and handlooms- and handicrafts-based. The policymaking and decision-making process has to be consciously decentralized and devolved to the regional level. Confidence-building measures and skill and capacity building projects are equally critical. These measures are required to enable the NER to lead India's Act East Policy initiative to its logical end.

INFRASTRUCTURE LAGGARDNESS

The infrastructure deficit remained a major physical constraint in the NER. Even today, the distance of the nearest railway station from the state capital varies from 0 km in Assam to 130 km in Mizoram and 216 km in Imphal. The distance from the state capital to the nearest airport varies from 5 km in Tripura and Manipur to 68 km in Arunachal Pradesh and 74 km in Nagaland. Of the total length of national highways of 13,658 km across the whole of the NER, Assam and Arunachal Pradesh constitute as high as 28 per cent and 18 per cent, respectively, as against the hardly 4 per cent of Sikkim. Further, of the total 2,830 km of railway lines, those in Assam constitute 90 per cent and those in Tripura 9 per cent, with literally negligible presence of railways in the other states. Despite a pan-India project like BharatNet connecting all the 11,252 *gram* panchayats in the NER, it continues to have the lowest bandwidth speed in India. As per the *Telecom Statistics 2018*, the teledensity in the NER is 64 per cent as against India's 83.4 per cent, and broadband and Internet connection are hardly 4.43 million and 11 million (2015–2016), respectively, which are about 3 per cent of national total. In 2018 more than 8,600 villages (22 % of total 40,000 villages) remained outside mobile connectivity (Department of Telecommunications 2018).

Over the last two decades, and more particularly in the last few years, the NER has seen a huge surge in physical infrastructure development and in the participation of newer players, like JICA and ADB. For instance, the NER constitutes over 10 per cent of

India's total roads spanning 46.90 lakh km and almost 14 per cent of the total national highways spanning 97,990 km. In the planned road projects, including the North–South and East–West corridor and two-lane roads to the district headquarters and backward and remote areas, 3,750 km has been completed. A total of 3,101 km of highways was built between 2014–2015 and 2018–2019 at the cost of ₹26,986 crore, and another 4,407 km is likely to be added with an investment of ₹47,476 crore.

A range of bridges, including the long-pending 4.94-km-long Bogibeel (₹5,900 crore land and rail), have been completed. All these projects, along with the upcoming mega projects, like the 1,850-km frontier highway in Arunachal Pradesh and new bridges over the Brahmaputra, would change the entire face of connectivity. Commissioning of and steady implementation of the 870.82-km railway lines connecting all the state capitals of the NER by 2022, at the cost of ₹42,239 crore, are underway. The 13 operational airports witnessed 98.4 lakh passengers and 38,094 tons of freight movement during 2018–2019.

Connectivity to operationalize India's Act East Policy has recorded massive progress with the alignment of Asian Highway 1 and Asian Highway 2 that connect India with Pakistan, Bangladesh and Myanmar and India with Nepal and Bangladesh, respectively, which pass through the NER. This includes the 1,360-km trilateral highway connecting Moreh (India), Mandalay and Yangon (Myanmar) and Mae Sot (Thailand) districts, the Kaladan Multi-Modal Transport Project connecting India and Myanmar (a 697-km waterway from Kolkata to Sittwe port in Myanmar and a 220-km roadway from Paletwa to the Indo-Myanmar border at Zorinpui), connectivity of India with Bangladesh through Sabroom in Tripura and Tura in Meghalaya and the proposed Bharatmala.

To effectively utilize the 3,839 km of total navigable waterways, a range of inland waterways, including the 891-km Sadiya–Dhubri (Bangladesh border) waterway under National Waterway 2 (NW-2) and the 121-km Lakhipur–Bhanga waterway, would constitute a new set of affordable transportation networks.

BORDERS AS OPPORTUNITY

Borders have constituted the key issues in interactions and exchanges with the neighbours of India. India's physical borders with its neighbouring countries were mostly drawn by Durand, McMahon and Radcliff hastily, crudely and unscientifically. The colonial legacy made the region think of the border purely as a geometric line. As borders became a bastion of orthodox military thinking, human-security concerns received little attention. Such borders have had far-reaching implications for this region as consistently manifested in protracted border-related conflicts, wars and political and diplomatic tug of war. The Act East policy gives us an opportunity to rethink, re-recognize and relocate these borders as borderlands wherein one can see the intrinsic interplay of natural resources, cultures, societies, trade–commerce, tourism, water towers, technology, roads and communication. There is a need to re-position the NER straddled by five international borders and with a strong historical narrative on the borderland. The moment we revive borderland ideas, borders become softer and interactive. The deeper elements of resilience and sustainability would be re-inducted, making economic integration and people-to-people contacts much more attractive and prolific (Lama et al. 2016).

In the borderland discourse, the core driving force has been connectivity. If India's burgeoning economic growth is to be shared by its neighbouring partners and if interdependence is to be consolidated, they would have to materialize primarily through a variety of cross-border physical, virtual and community linkages. Through the transformation of physical borders into smart borders, the borderlands between China and Southeast Asian countries became locations for vibrant economic activities. Similarly, when India finally allows Bangladesh, Bhutan and Nepal to take advantage of the India–Myanmar–Thailand Trilateral Highway under its Act East policy and when the BBIN and BCIM[12] come into fruition, the NER with its historic cross-border connections would become the central force of gravity, putting an end to the 'connectivity black hole' syndrome that affected the NER all these decades.

Similarly, in all three conspicuously attractive trans-border projects—the China–Pakistan Economic Corridor (CPEC) that links Gwadar Port with Karakoram Highway and the western topography of China, finally leading to integration with Central Asia, the India–Afghanistan–Iran project at Chabahar in Iran and the ongoing interconnection between Nepal's Rasuwagadhi and China's Geirung (Kerung) through Shanghai–Lhasa–Shigatse railway lines and highways—there are clear indications that borderlands and not borders are going to be the theme of engagements. India's experience in cross-border interconnections and the initiatives such as the reopening of the once-versatile Nathu La trade route between Sikkim and Tibet Autonomous Region after 44 years in 2006 partly convince the sub-region about the prosperous criticality of reconnecting with the borderlands. Today, the Nathu La route, with an annual trade exchange of ₹48.28 crore in 2018–2019, is widely used for the Kailash–Mansarovar pilgrimage yatra.[13]

DEVELOPMENT PROSPECTS IN THE NORTH EASTERN REGION: BASIC QUESTIONS

Despite the unfurling of a series of developments over the last decade or so and Act East becoming a flagship foreign policy agenda for the Government of India, there are questions that are frequently and vociferously raised. These are primarily related to the re-positioning of the NER in India's larger national security, foreign policy and development dynamics.

Where does the NER figure in the Act East project of India? Why has there been no clear official enunciation as yet by the Government of India on the Act (Look) East policy and the role of the NER regarding which deliberations could be done, strategies could be designed and concrete actions could be initiated?

How does one reorient the economies of the NER? What interventions could bring a new paradigm of development where both domestic and foreign investments and other sectoral participation converge into this region? Could some of the investment and industrial projects of Japan, particularly the *traditional middle- and low-end industries that are now shifting base from China, be inducted into this region?*

What are the critical institutional interventions required? What are the nuances of the 'cooperative federalism' and 'neighbourhood first policy' of the present NDA government which need to be practised in the NER? Could the 'sub-state diplomacy' and 're-imagined borders' strategies extend 'full-stakeholder' status to the NER?

Given that cooperation is a two-way practice, how do the Association of South East Asian Nations (ASEAN) member countries perceive this Indian initiative? Do they have the political commitment and the requisite operational and institutional capabilities to move towards the West and finally integrate with the NER and rest of India?

What are the core geopolitical issues?
What have been the experiences of other sub-regional groupings that involve China and India's immediate neighbouring countries, like ASEAN countries?

How could the issues of energy, climate change, traditional knowledge, reopening of old roads and passes, war residues tourism or institutionalized gene pool marketing be used as critical variables in the trans-border multi-modal corridors?

THE NORTH EASTERN REGION AS A LEADER

Under the new mode of reverse integration, the NER could take the lead in some crucial areas where it has a distinct comparative advantage. For instance, as a gateway and multi-modal corridor to Southeast Asia for both India and its immediate eastern neighbours, the NER, along with Bhutan, Bangladesh, Myanmar and Nepal, could actually be the power/electricity pool of South Asia in the eastern junction. The surplus electricity generated in Bhutan, Nepal and the NER could be pooled together to a sub-regional pool in the NER and distributed through cross-border transmissions to Southeast Asia. The power exchange between Tripura in the NER and Bangladesh triggered by the 726-MW combined cycle gas turbine (CCGT) at Palatana (Tripura) provides a new direction in terms of local integrative exchange. Besides catering to areas of the NER facing a power deficit, this project exports 100 MW to Bangladesh (export of 100 MW more is being agreed to by the Tripura government), mainly in lieu of the services provided by the latter in transporting the project-related equipment and goods and services through its waterways via Kolkata. About 2–3 MW is exported to Tamu town in Myanmar through a 11-kV transmission line from Moreh in Manipur. A new study on energy in India's Act East policy and the role of the NER by Economic Research Institute for ASEAN and East Asia (ERIA) in Jakarta shows how the NER could be a critical player in the energy trading game in this sub-region (Anbumozhi et al. 2019).

For the NER to be a pivot in the Act East policy, it has to design and plan a four-way comprehensive connectography, namely, within states, between the states in the NER, with the rest of India and with the neighbouring countries. This quadrilateralism very much fits into what Prime Minister Modi has been advocating in terms of connecting the entire South Asian region by HIT, that is, highways, information technology and transmission lines. Further, if the NER has to be a gateway to the East, which used to be the case till 1947, there have to be conscious and time-bound efforts to address and resolve major gaps, like (a) knowledge and information gaps; (b) seed projects and start-up gaps; (c) matchmaking, investment, technical facilitation, human resources and capacity gaps; (d) policy coordination and institutional harmonization gaps; (e) confidence gaps; and (f) technology gaps. These gaps are rampant across the NER and have prevented any meaningful economic actors and investors, both from within and outside India, from taking part in sustained development projects.

CHINA'S DIVERSION OF RIVERS: IMPLICATIONS FOR THE NORTH EASTERN REGION

The impact of climate change is very visible in the NER. Warm winters, rhododendrons flowering in January instead of in March–April, loss of full spring season, forest fires, ecological surprises and changes in rainfall pattern affecting the entire hydrological flows in the rivers, glacial erosions, the visible phenomenon of phenology and new varieties of human, plant and animal diseases and severely damaging natural disasters have become regular occurrences. The people are worried on the water security front. For instance, the water sources like that in Tibetan and Qinghai plateau in China have been impacted upon with adverse consequences on hydrological flows and traditional spatial dynamics in the NER. Three adverse and deleterious dimensions are emerging. The mountainous NER has witnessed steady drying up of springs (Lama 2001) and the age-old river systems. The east-flowing rivers in China, like the Brahmaputra, have recorded a sharp decline in environmental flows, as the tributaries themselves are fast drying up in Tibet.[14] Along with the dams for power projects and other barrages and flow barriers on rivers flowing down to the NER (Bandyopadhyay et al. 2016; Shah and Giordano 2013; Siddiqui et al. 2018), a substantive intervention on river diversion has been initiated mainly to cater to the increasing alarmingly dry regions of China (Lama 2018; Ma Jun 2004; Merta 2010; Tan et al. 2015). Government of India should intervene and negotiate with China to ensure both hydrological and environmental flows, which are the lifelines of the communities living in the Northeast.

INSTITUTIONAL FEDERALISM

The NER has a lot to learn from the contributions of M. S. Swaminathan, the father of the Green Revolution, and Verghese Kurien, who headed the National Dairy Development Board (NDDB). Both of them had an incredibly deep conviction about the value of institutions and their unenviable and priceless role in bringing communities, societies, geographies and countries together. Being far away from Delhi can be a blessing in disguise, as it enables establishing new institutions and experimenting with new opportunities. In a way, China has shown how even landlocked and peripheral regions can grow and prosper. As early as the 1990s, China consciously devised ways and means to give leeway and autonomy to its border provinces to engage with the immediate neighbourhood. The initiatives taken by Jilin, Yunnan and Tibet to connect with neighbouring countries epitomize the 'acting federalism' of China (Freeman 2011; Golley 2007).

A diverse country like India requires institutional federalism and functional autonomy for various institutions. Functional autonomy for the NER would enable the states to explore alternative possibilities of growth through strengthening their relations with neighbouring countries. The former Vice President Hamid Ansari terms this as 'Sub-State diplomacy', and former Foreign Secretary Shyam Saran calls for 'reimagined borders'. The prime minister's project of 'cooperative federalism' could address some of these crucial questions and be an instrument in the process of reverse integration with the NER. The inclusion of the chief ministers of the NER in the prime ministerial delegation to ASEAN countries and Bangladesh and mandating of visits by the central ministers to the NER are healthy indicators of changing nuances of cooperative federalism.

It is time for the NER to create an operational architecture of its own within the framework of cooperative federalism. A high-level 'Act East Forum' led by the prime minister should be set up where, besides the ministers and secretaries of all the concerned line ministries in New Delhi, chief ministers of the NER and West Bengal and their chief secretaries, members of the NITI Aayog and the North Eastern Council, all the ambassadors of India located in ASEAN and East Asian countries and academics and civil society actors could be members. The national security adviser and the foreign secretary could play pivotal roles in their respective domains. The chief ministers of the NER may be bestowed with target-oriented roles in some major areas, including in Road and Transport Mission, Energy Mission, Horticulture and Flori-culture Mission, Trade Mission, Investment Mission, Technology Mission, Plantation Mission, Tourism Mission, Education Mission, Youth Mission, etc.

The MDoNER could now be in the NER and headed by a union cabinet minister. This would also require drastic reorientation and real-location of functions. The new MDoNER would have representatives of all the key line ministries of the union government in its secretariat at the joint secretary/additional-secretary level. These would include finance, rural development, health, education, planning, culture, forest, environment, roads and communications, foreign affairs, home, defence, shipping and transport. The revamped MDoNER could have development commissioners of states, eminent civil society members and private-sector players who would be responsible for operationalizing the crucial aspects of the Act East policy. It should become the secretariat and operational agency for projects within the NER and which relate to varieties of flows across the border and frontier-zone engagements with Southeast Asian countries. The North Eastern Council would be a unit of the MDoNER and could be transformed into a counterpart of the NITI Aayog, assigned the roles of planning, monitoring and evaluation. This would also require drastic restructuring of the North Eastern Council as a regional planning body through induction of professionals.

Further, it would make sense to make the Darjeeling and Dooars region of West Bengal the North Eastern Council's ninth member, as this region has natural contiguity and sociocultural similarities with the NER and acts as the actual gateway between the NER and the rest of India. This would also provide the much-required geographic compactness to the NER and bring about durable peace and stability in the conflict-prone Darjeeling district. A 10-year road map may be drawn by the MDoNER based on a sufficient number of doable projects identified in a series of reports, including the Vision Document 2020, and also projects identified and agreed upon by India and various countries in ASEAN, including in infrastructure, health, trade, investment, security, education, tourism and agriculture. The ASEAN and East Asian countries could have their consulates in the NER. Initiation of multi-layer interdisciplinary dialogues and consultations among various stakeholders, including universities and think tanks and development partners both within and outside India, would have far-reaching effects on the future of the NER.

NOTES

1. To be incorporated in the category of special-category states, five inherent characteristics are required: (a) hilly and difficult terrain; (b) low population density or sizeable share of tribal population; (c) strategic location along borders with neighbouring countries; (d) economic and infrastructural backwardness; and (e) non-viable nature of state finances. Under the Normal Central Assistance earmarked by the Planning Commission, the special-category states receive 90 per cent grants and 10 per cent loans (90:10), while the same ratio for other states has been 30:70. See PRS Legislative Research (2013).

2. The projects taken up under the Technical Cooperation (TC) project of Japan international Cooperation Agency (JICA) include Capacity Enhancement for Sustainable Agriculture and Irrigation Development in Mizoram, North East Road Network Connectivity Improvement Project, Sustainable Catchment Forest Management and Continuous Improvement of Forests in Tripura. See Japan International Cooperation Agency, https://www.jica.go.jp/india/english/office/topics/press170807.html; https://www.jica.go.jp/english/our_work/social_environmental/id/asia/south/india/c8h0vm0000egxyoc.html; https://www.jica.go.jp/india/english/office/topics/press190116.html.

3. The International Fund for Agricultural Development (IFAD) has now under-taken the US$30 million second phase of North Eastern Region Community Resource Management Project for Upland Areas (NERCORMP) covering several districts in Assam, Meghalaya, Arunachal Pradesh and Manipur. See *Business World* (2020).

4. For instance, the World Bank has undertaken the US$130 million North East Rural Livelihood Project for India to improve rural livelihoods, especially those of women, unemployed youths and the most disadvantaged, in the participating north-eastern states. It has been implemented by the Ministry of Development of North Eastern Region (MDoNER) in 11 districts in Mizoram, Nagaland, Sikkim and Tripura. See World Bank (2011).

5. The Asian Development Bank (ADB) has undertaken the MFF—North Eastern State Roads Investment Program to: (a) improve about 430 km of priority road sections in six states (Assam, Manipur, Meghalaya, Mizoram, Sikkim and Tripura) in the North Eastern Region (NER) of India; and (b) provide capacity building support to the executing agencies—the MDoNER and the state public works departments (PWDs) or their equivalent—in the six project states. The investment programme will target the secondary road network and aims to enhance the performance of the state roads sector in the NER through investment project implementation and dedicated capacity building measures. The improved secondary road network would provide important linkage between the primary and tertiary road networks in the region, for which there are ongoing national programmes for improvement. For details, see Asian Development Bank (2013).

6. See Planning Commission, Government of India (1997); Ministry of Finance, Government of India (2013); 'National Commission to Review the Working of the Constitution', Advisory Panel on Decentralisation and Devolution; Empowerment and Strengthening of Panchayat raj Institutions, 2001; Planning Commission, Government of India (2006); Reserve Bank of India (2006); Report of the Expert Committee, Ministry of Panchayati Raj, Government of India (2006); Lama (2001); Lama et al. (2008); World Bank (2007); and Lama (2005).

7. For instance, the distance from Agartala to Calcutta before the partition was only 350 km, but now through the Siliguri corridor it is 1,645 km. See Schendel (2005).

8. For details of the outcomes of the relevant meeting of the Union Minister of State (Independent Charge) of the MDoNER, see http://pib.nic.in/newsite/PrintRelease.aspx?relid=114721

9. In the letter, Sardar Patel wrote: 'The people inhabiting these portions have no established loyalty or devotion to India. Even Darjeeling and Kalimpong areas are not free from pro-Mongoloid prejudices. During the last three years, we have not been able to make any appreciable approaches to the Nagas and other hill tribes in Assam'. See http://www.darjeeling-unlimited.com/patel.html

10. To mention a few, II World War, the Sino-Indian War of 1962, the freedom movement in Bangladesh, the Naga/Mizo/Manipur/Tripura insurgencies, the Naxalite movement, merger of Sikkim and India, movements for Gorkhaland, Bodoland, etc., students' movements against foreigners in Assam, refugee exodus from Bhutan, Myanmar, Tibet and Nepal and conflicts over water sharing with neighbouring countries.

11. Some sportspersons from across the NER who have started bringing laurels to India are Mary Kom (boxing), Anshu Jamsenpa (mountaineering), Hima Das (athletics), Anuradha Devi Thokchom (hockey), Bombayla Devi Laishram (archery), Chekrovolu Swuro (archery), Dipa Karmakar (gymnastics), Tarundeep Rai (archery), Shiva Thapa (boxing), Gohela Boro (archery), H. Lal Ruat Feli (hockey), Kalpana Devi (judo), Ngangbam Soniya Chanu (weightlifting), Saikhom Mirabai Chanu (weightlifting), Sushila Chanu (hockey) and Bhaichung Bhutia and Sunil Chhetri (football).

12. For details of the joint statement of India and China during Prime Minister Narendra Modi's visit to China, issued on 15 May 2015, see http://www.mea.gov.in/bilateral-documents.htm ? dtl/25240/Joint_Statement_between_the_India_and_China_during_Prime Ministers_visit_to_China) (accessed 4 August 2015).

13. This author happened to be the Chief Economic Adviser in the Government of Sikkim (2002–2007) and was assigned the task of writing the policy report by the government, mainly focusing on the modalities, implications and prospects of reopening this historic trade route. The report writing team consisted of well-known experts on national security, international trade and investment, connectivity, cultural history and other related issues. The team visited a range of border crossing points, including those in South Asia, China, Southeast Asia and Japan. It made a comprehensive market survey in Lhasa city and around. See Lama (2005).

14. One witnesses a number of dried tributaries of Yarlung Tsangpo with pebbles and sand totally exposed while travelling through the Gyantse–Shigatse–Saga–Tharche highway on the Manasarovar and Kailash route in Tibet Autonomous Region of China. Despite this, the hydrological flows of Yarlung Tsangpo are quite handsome in a city like Lhasa (observation of this author after accomplishing the Kailash–Manasarovar yatra).

REFERENCES

Anbumozhi, Venkatachalam, Ichiro Kutani, and Mahendra P. Lama. 2019. *Energising Connectivity Between Northeast India and Its Neighbours.* Jakarta: Economic Research Institute for ASEAN and East Asia.

Asian Development Bank. 2013. 'India: MFF—North Eastern State Roads Investment Program (Facility Concept)'. https://www.adb.org/projects/37143-013/main (accessed 8 February 2021).

Bandyopadhyay, Jayanta, Nilanjan Ghosh, and Chandan Mahanta. 2016. *IRBM for Brahmaputra Sub-basin Water Governance, Environmental Security and Human Well-being*. New Delhi: Observer Research Foundation.

Bhaumik, Subir. 2009. *Troubled Periphery: Crisis of India's North East*. New Delhi: SAGE Publications.

Business World. 2020. 'IFAD Approves USD 30 Million For North East Project'. *Business World*, October 5. http://www.businessworld.in/article/IFAD-Approves-USD-30-Million-For-North-East-Project/09-11-2017-131236/ (accessed 8 February 2021).

Centre for Studies in International Relations and Development. 2009. *BIMSTEC Cooperation Report 2008*. Kolkata: Bookwell Publications.

Freeman, Carla Park. 2011. *China on the Edge China's Border Provinces and Chinese Security Policy*. Washington, DC: Johns Hopkins University School of Advanced International Studies.

Golley, Jane. 2007. *The Dynamics of Chinese Regional Development*. Cheltenham: Edward Elgar.

Government of India. 2004. *Report of the Committee on Revitalization of the North Eastern Council*. New Delhi: Ministry of Development of North Eastern Region.

Government of India. 2006. 'Planning for the Sixth Schedule Areas and those areas not covered by Parts IX and IX-A of the Constitution'. Report of the Expert Committee, Ministry of Panchayati Raj. New Delhi: Government of India.

Hettne, Bjorn. 1999. 'Globalisation and the New Regionalism: The Second Great Transformation'. In *Globalism and the New Regionalism*, edited by Bjorn Hettne, 1–24. London: Macmillan.

Lama, Mahendra P. 1989. *The First Plan of Darjeeling Gorkha Hill Council*. Darjeeling: DGHC.

Lama, Mahendra P. 2001. *Sikkim Human Development Report*. New Delhi: Social Sciences Press.

Lama, Mahendra P. 2005. *Sikkim-Tibet Trade via Nathu la: A Policy Study on Prospects, Opportunities and Requisite Preparedness*. Gangtok, Sikkim: Nathu la Trade Study Group for the Government of Sikkim.

Lama, Mahendra P. 2018. 'River Diversion in China'. *Kathmandu Post*, February 14, 2018. http://kathmandupost.ekantipur.com/news/2018-02-14/river-diversion-in-china.html .

Lama, Mahendra P. et al. 2008. *Sikkim Development Report*. New Delhi: Planning Commission, Government of India, Academic Foundation.

Lama, Mahendra P., Ling, L. H. M., Banerjee, P., Abdenur, A. E., Kurian, N., & Li, B. 2016. *India and China: Rethinking Borders and Security*. Michigan: University of Michigan Press.

Ma Jun. 2004. *China's Water Crisis*. Norwalk: East Bridge Books.

Mertha, Andrew C. 2010. *China's Water Warriors: Citizen Action and Policy Change*. New York: Cornell University Press.

Ministry of Finance, Government of India. 2013. 'Report of the Committee for Evolving a Composite development Index of States'. September. https://finmin.

nic.in/sites/default/files/Report_CompDevState.pdf (accessed 8 February 2021).

Nath, Hiranya K. 2005. 'The Rise of Enclave Economy'. In *India's North East: Development Issues in a Historical Perspective*, edited by Alokesh Barua, 1–21. Delhi: Manohar.

NITI Aayog. 2015. *Evaluation Study on Border Area Development Programme (BADP)*. PEO Report No 229. New Delhi: Government of India.

Paul, Bappaditya. 2014. *The First Naxal: An Authorised Biography of Kanu Sanyal*. New Delhi: SAGE Publications.

Pemberton, R. B. 2015. *The Eastern Frontier of India (1835)*. New Delhi: Mittal Publications.

Planning Commission, Government of India. 1997. 'Transforming the North East: Tackling Backlogs in Basic Minimum Services and Infrastructural Needs'. High Level Commission Report to the Prime Minster.

Planning Commission, Government of India. 2006. 'Report of the Task Force on Connectivity and Promotion of Trade and Investment in North East States'. http://megplanning.gov.in/report/Task_Force_Report.pdf (accessed 8 February 2021).

PRS Legislative Research. 2013. 'Special Category Status and Centre-State Finances'. https://www.prsindia.org/theprsblog/special-category-status-and-centre-state-finances (accessed 8 February 2021).

Rahman, Musfizur, Habibur Rahman, and Wasel Bin Shadat. 2007. *BCIM Economic Cooperation; Prospects and Challenges*. Occasional Paper 64. Dhaka: Centre for Policy Dialogue.

Reserve Bank of India. 2006, July. 'Report of the Committee on Financial Sector Plan for North Eastern Region'. Mumbai: Reserve Bank of India.

Sarkar, R. L., and Mahendra P. Lama. 1986. *The Eastern Himalayas: Environment and Economy*. Delhi: Atma Ram & Sons.

Schendel, Willem van. 2005. *The Bengal Borderland: Beyond State and Nation in South Asia*. London: Anthem Press.

Shah, Tushaar, and Mark Giordano. 2013. 'Himalayan Water Security: A South Asian Perspective'. *Asia Policy* 16:26–31.

Siddiqui, Shawahiq, Shilpa Chohan, and Partha Jyoti Das. 2018. *Reimagining Brahmaputra Policy and Regulatory Aspects of Transboundary Water Governance: A Scoping Study*. New Delhi: OXFAM India.

Tan, Debra, Feng Hu, and Hubert Thieriot. 2015. *Towards a Water & Energy Secure China: Tough Choices Ahead in Power Expansion with Limited Water Resources*. China Water Risk. https://www.chinafile.com/library/reports/towards-water-energy-secure-china (accessed 8 February 2021).

Thavaraj, M. J. K. 1972. 'Regional Imbalances and Public Investment in India (1860–1947)'. *Social Scientist* 1(4):3–24.

Weiner, Myron. 1993. 'Security, Stability and International Migration'. In *International Migration and Security*, edited by Myron Weiner, 91–126. Oxford: West View Press.

World Bank. 2007, June. *India: Development and Growth in the North East India: The Natural Resources, Water and Environment Nexus, Strategy Report*, 23. Washington: World Bank.

———. 2011. 'Restructuring Paper on A Proposed Project Restructuring of North East Rural Livelihoods Project (NERLP) Approved On December 20, 2011 to Ministry Of Development of North East Region'. Report RES35908. http://documents.worldbank.org/curated/en/712261551473678421/pdf/Disclosable-Restructuring-Paper-North-East-Rural-Livelihoods-Project-NERLP-P102330.pdf (accessed 8 February 2021).

Chapter 2

Political Economy of Road Infrastructure in India's Northeast since Liberalization

Rakhee Bhattacharya

Over the 19th and early 20th centuries, road-building projects in the north-eastern province/frontier of Bengal constituted one of the essential imperial undertakings. The imperial routes and transport lines primarily were propelled through the undivided Bengal province, to secure access to the adjoining areas and to connect the region to the global trade centres. This dire need for an improved road system subsequently compelled the imperials to establish a Public Works Department (PWD) in 1868 under a separate North-East Frontier Tracts administration system. By 1937–1938, the imperials built about an 8,000-km-long road network to provide access to the jungle-laden, flood-prone and mountainous frontiers of the Northeast (Das 2009; Ganguly 2006; Sharma 2011). However, this 'reworlding' of the frontier with the imperial road transport network could neither provide any long-standing territorial identity nor prove the conventional wisdom to bestow any economic prosperity (Demenge 2015; Soja 2000). Roads were primarily used for extraction, appropriation and transportation of the vast resources of the Northeast to the making of the imperial metropolis.

This geo-history of the Northeast frontier subsequently was 'uprooted' in the post-independence period. The long-drawn-out cartographic exercises produced about 4,000 km of international borders with a new imagination of the Northeast as a borderland (Pachuau and Schendel 2016; Rustomji 1983; Soja 2000). These borders produced multi-layered deadlocks and infrastructural asymmetry, affecting the mobility, connections and everyday life of the Northeast people, and they shaped a set of complex external orientations with the neighbouring nations of East Pakistan (Bangladesh), Burma (Myanmar), Tibetan Region (of China), Nepal and Bhutan. With natural neighbours turning into 'strategic neighbouring nations', many of the earlier imperial cross-border roads of the Northeast were rendered inactive. New road-building projects were subsequently halted, pushing the space expediently to the margins and representing it as geographically 'cut off'. Only a 28-km-long stretch of a road corridor through Siliguri town of North Bengal, commonly known as the Chicken's Neck, could sustain the physical connectivity of the Northeast with other parts of India, depicting its external imagination and representations as a 'remote periphery'. Subsequently, in 1971, with the passing of the North-Eastern Areas (Reorganisation) Act, the states of the Northeast were dropped from the eastern zone of India, with a concerted effort to make this an 'economic zone' (Chaube et al. 1975). Confronted with an acute challenge of connectivity infrastructure, the fundamental criterion for economic development, a regional planning institute, the North Eastern Council (NEC), was created in the following year, 1972, to undertake new road-building projects across the hills and valleys of the north-eastern states. This integrated approach to road infrastructure was meant to connect this geographically 'cut-off' region of India, and in the next 40 years, as the NEC claimed, about 9,800 km of roads and 77 new bridges were built within the region (Bhattacharya 2019). However, with a large number of them remaining incomplete, poorly maintained and subsequently dilapidated, the region continued to face a severe road infrastructure deficit and remained a 'remote periphery' of India.

A series of new policy reforms however unfolded during India's liberalization programme to bridge this acute connectivity–infrastructural gap in the Northeast. New attempts were made to change the narrative of the Northeast with India's structural reforms in its economic policy from a nationalistic frame to a neoliberal approach. With the changing spatial and temporal specificities, road-building in the country beyond the core metropolis to various remote and far-flung peripheries was justified. With the expectation of better economic prosperity with the liberalization programme, India introduced the Look East policy in the early 1990s. This was to extend economic engagement beyond the borders to the neighbouring nations of Southeast Asia, and the spatiality of India's remote north-eastern periphery was re-imagined as a 'connected gateway' because of its geographical proximity to those neighbouring nations. In this environment, a robust highway infrastructure system across the harsh, hilly and meandering geography of the Northeast was a pre-eminent means to achieve such transnational economic engagements. Highway infrastructure in the Northeast was invoked as the key for local and regional economic development like elsewhere in the world (Bourguignon and Pleskovic 2008; Fujita et al. 1999; Scott 1998; Storper 1997; World Bank 1994). Over the years, as this global programme got embedded in India's policy reforms, a more aggressive Act East policy was invoked in 2014 to re-emphasize strengthening and extending the regional highway infrastructure of the Northeast. It was identified as a means to achieve 'inclusive growth' through connecting to such isolated and landlocked areas, similar to the global trend (Perz, Xia, and Shenkin 2014).With geographical significance, the Northeast naturally gained more attention in the policy space to become a potential 'economic corridor' for India to connect with the Asia-Pacific region (Anbumozhi et al. 2019; Bhattacharya 2019; De et al. 2019; Yumnam 2019).

HIGHWAY PROJECTS IN THE NORTHEAST AND THE NEOLIBERAL STATE POLICY

In the year 1998, the National Highways Development Project (NHDP) was introduced by the National Highways Authority of India (NHAI) as an administrative unit for executing new highway projects to connect all the core metropolises of Delhi, Mumbai, Chennai and Kolkata with four-lane and six-lane National Highways (NHs). It became a part of India's ambitious Golden Quadrilateral (GQ) project for achieving an expansive and integrated economy. Some recent empirical studies have shown that the GQ project made an impact on India's economy, including the manufacturing sector (Datta 2012; Ghani et al. 2016). The North–South and East–West Corridor (NS-EW) was the other big highway project of the NHAI, which was to connect the furthest peripheries of Kashmir and Kanyakumari and Silchar and Porbandar. The far-flung and bordering areas of the Northeast, for the first time in the post-independence period, became part of such ambitious highway infra-link projects of India. Subsequently, in 2006, a unique central flagship scheme, Special Accelerated Road Development Programme in the North East (SARDP-NE), was introduced to build highway networks in this region. It aimed to connect all the 88 district headquarters of the states of the Northeast and envisaged bridging the 'yawning infrastructural gap' that had been created during the post-independence period through building both two-lane and four-lane highways and also through reaching up to the toughest mountainous bordering areas. The programme was implemented in two phases with an exclusive Arunachal Pradesh package to build a 10,141-km-long highway network. The construction was divided into two phases, with Phase A having 4,099 km of roads, with 2,319 km only for Arunachal Pradesh. In Phase B the remaining 3,723 km was constructed. Of this total length, 47 per cent (4,798 km) was accounted for by national highways, and the remaining 53 per cent (5,343 km) by state highways. This was a historic policy step that essentially had multidimensional objectives, like changing the narrative of inaccessibility of the Northeast, achieving new goals of economic engagements with neighbouring nations, ensuring economic and investment opportunities for this 'underdeveloped region' and improving the well-being of the border communities at the local levels. The highway infrastructure also was meant to facilitate India's defence logistics across the bordering areas.

A look at the data of the Ministry of Road Transport and Highways (MoRTH) for the period 2005–2017, shown in Table 2.1, shows that the region is now at par with the national level in terms of percentage share of highways in the total road networks. All the states of the Northeast have had a significant increase in both national and state highway networks, reversing the role of the post-independence state of India, which had critically peripherialized this region in its

Table 2.1 *Highway Networks in the Northeast: Between 2005 and 2017 (km)*

States	National Highways		State Highways		Total Road Networks	
	2005	2017	2005	2017	2005	2017
Arunachal Pradesh	392 (2.10)	2,513 (6.79)	0	8,123 (21.94)	18,690	37,025
Assam	2,836 (1.35)	3,844 (1.14)	2,781 (1.33)	2,530 (0.75)	209,591	337,777
Manipur	959 (5.68)	1,745 (6.32)	1,118 (6.62)	715 (2.59)	16,897	27,612
Meghalaya	810 (7.90)	1,203 (5.24)	1,134 (11.06)	772 (3.37)	10,249	22,939
Mizoram	927 (14.90)	1,382 (12.55)	140 (2.25)	170 (1.54)	6,220	11,012
Nagaland	494 (1.80)	1,173 (3.24)	404 (1.47)	722 (1.99)	27,459	36,239
Sikkim	62 (1.840)	463 (4.07)	186 (5.52)	663 (5.82)	3,371	11,386
Tripura	400 (1.24)	806 (1.88)	689 (2.14)	329 (0.77)	32,149	42,925
NER	6,880 (2.12)	13,129 (2.49)	6,452 (1.99)	14,024 (2.66)	324,607	526,915
All India	65,569 (2.25)	114,158 (2.28)	144,396 (4.96)	175,036 (3.50)	2,909,156	4,997,671

Source: Calculated from Basic Road Statistics of the Ministry of Road Transport and Highway (MoRTH).
Note: Figures in parentheses are percentages of the total road.

national road-building programme. The SADRP-NE has been an impactful post-liberalization policy intervention to introduce new highway networks in this region and to reverse the narrative of the post-independence period. However, it has remained unsuccessful in ensuring uniform road infrastructure across the states of the region, and substantive intra-regional variations exist, with some states of the Northeast achieving better results than others. Arunachal Pradesh has achieved the most. The ambitious Trans-Arunachal Highway project was undertaken to make this remotest mountainous and bordering state accessible through a 1,559-km east–west highway stretch, through connecting 12 out of its 16 districts. This was primarily a geopolitical strategy to counter the neighbouring China's road infrastructure in the Tibetan region.

The rising significance of highway infrastructure in this strategic region, along with increasing expenditure, pushed a new central administrative authority, National Highways and Infrastructure Development Corporation Limited (NHIDCL), in 2014. The NHIDCL is meant to develop and improve highways of 10,000 km length in the region and facilitate the creation of international trade corridors through the Northeast. An additional 500 km is also planned to expand the regional transport network through connecting North Bengal, the Northeast and the member nations of the South Asia Subregional Economic Cooperation (SASEC). This highway expansion is to remove the decade-old narrative of the Chicken's Neck, the sole connecting corridor of the Northeast, and categorically to promote cross-border trade and commerce across the eastern Himalayan region, and also to safeguard India's international borders (NHIDCL 2018). It has generated a total capital corpus of ₹1.41 lakh crore for various civil works of such ambitious highway projects in the region, while an additional ₹1.07 lakh crore has been targeted for the costs related to various land acquisition and rehabilitation processes. A fresh survey has also been undertaken to design, build, operate, maintain and upgrade all National Highways and strategic roads in the Northeast through sharing a large amount of technical know-how and enhancing relations and opportunities for business development in the Northeast. This idea of efficient highway logistics in the region also unfolds the role of an authoritative state that aims for economic gains and trade surplus with the neighbouring nations through appropriating and extracting the resources of the Northeast.

This regained emphasis pushed the state to create another unique road project for this region, the North East Road Sector Development Scheme (NERSDS), in 2015–2016. This was to focus more on the interstate road network and to address the intra-regional road asymmetry, while ensuring various strategic connections. Two strategic bridges were completed in 2017 and 2018—the 9.1-km-long Dhola–Sadiya and the 4.9-km-long Bogibeel bridges—which now connect the remotest Arunachal Pradesh with Assam, and then to the other states. As Arunachal Pradesh had been the most 'cut-off' state for several decades, these two bridges brought immense relief to the local communities. An ethnographic study was conducted in 2018, which revealed that people were happy with these two bridges, especially in the eastern part of Arunachal Pradesh. The bridges have largely removed the geographical isolation of Arunachal Pradesh and subsequently increased travel and tourism. These bridges also facilitated easy military movement to the eastern part of Arunachal Pradesh up to Walong, the bordering town of the state, and helped secure India's position against the aggressive China in that vulnerable area. Forty-seven more such economically and strategically important but neglected interstate roads in the Northeast are being identified, with a total length of 1,665.75 km. These are seen as crucial for local mobility and connections and would help change the lives of the people through creating both forward and backward linkages and providing access to the markets. With the new improved roads, the existing old and dilapidated interstate roads, mostly constructed and maintained by the NEC and Border Roads Organisation (BRO), have also been categorized as 'Orphan Roads' for restoration and upgradation under the NERSDS.

Other major highway projects that are cross-border in nature, such as the 1,360-km Trilateral Highway to connect India, Myanmar and Thailand via Manipur, the Kaladan Multi-Modal Transit Transport Project to connect Kolkata and Myanmar via Mizoram, the EW to connect Gujarat and Southeast Asia via Silchar and then Manipur as part of the Mekong–India Economic Corridor (MIEC), are fast changing the external imagination of the Northeast as a 'connected geography'. An additional 24 such cross-border connectivity projects with Bangladesh, Bhutan, Nepal and Association of South East Asian Nations (ASEAN) countries have been undertaken in the last 5 years, some of which are complete now, especially those with Bangladesh through Tripura and Meghalaya. Both these interstate and cross-border highway infrastructures are largely meant to 'transform' the Northeast into an 'aspirational' region of India. They are expected to create many production hubs through exploring the vast resources of the region. This is also expected to create opportunities and ensure the well-being of the people with new livelihood opportunities. The immense gap in infrastructure and the economic 'underdevelopment' in the past had created a low-level equilibrium trap in the region with multiple human security challenges. For example, based on the latest National Sample Survey Office (NSSO) data, poverty ratio in the Northeast was 30.89 per cent, which was higher than India's 25.7 per cent (Bhattacharya and Bhattacharjee 2018). This poverty ratio also has huge intra-regional variation and is the worst in the states like Manipur and Assam. Some NSSO regions of these two states showed poverty ratios ranging between 50 and 80 per cent (Bhattacharya and Bhattacharjee 2015). The increase in the size of the formal economy of the Northeast during the period between 2004–2005 and 2015–2016 could not ensure effective redistribution and welfare policies (according to CSO data, the formal economy of the Northeast had grown from ₹96,855 lakh in 2004–2005 to ₹378,221 lakh in 2015–2016). Nor could the highway infrastructure make any meaningful impact on the lives of the people, as they are trapped in such high poverty. The region is also trapped with high unemployment rates. The latest Periodic Labour Force Survey report of 2017–2018 shows that most of the north-eastern states have higher unemployment rates than the national average of 17.8 per cent. It is the worst among the age group of 15–29. Nagaland tops the list in India with 56 per cent youth unemployment, while Manipur, Mizoram, Assam, Arunachal Pradesh and Tripura have 35.7 per cent, 28.6 per cent, 27 per cent, 26 per cent and 19.9 per cent youth unemployment respectively. Lack of enough economic opportunities forces high competition for government jobs, where ethnic composition plays a dominant role, and 'access remains synonymous with social mobility for a large section of the population'. The rest are pushed out of the region or are engaged in various unorganized and informal economies. Organized industry, which is a channel to absorb additional workforce, has merely had any presence in the region. During the period from 2004–2005 to 2015–2016, the number of industrial units increased from 2,211 to 5,096. The number of people engaged in these industries, however, increased from 142,633 to 278,149 only (Bhattacharjee and Bhattacharya 2018). The SARDP-NE that was expected to improve economic activities to create livelihood opportunities for the people of the region has failed to achieve its objective. However, this has been an effective policy in the post–liberalization period in transforming the narrative of the Northeast from that of a 'cut-off region' to that of a 'connected gateway' to the larger political and economic representations of India and its neighbouring nations and in ensuring a modern territorial identity for the Northeast.

ROAD PROJECTS AND THE CAPITAL REGIMES IN THE NORTHEAST

Traditionally, the state is responsible to create infrastructure logistics for the citizens with large-scale and long-term fiscal outlays. Such high state budgetary resources create a deep dynamic of power relations, strengthen monopoly of state capitalism and reinforce bureaucratic influence in the process of infrastructure creation. The highway infrastructure in the Northeast had similarly pushed huge central fiscal outlays, and the SARDP-NE alone was initiated with an estimated corpus of ₹40,000 crore, which at the beginning was mostly raised through various state heads, such as through the Pradhan Mantri Gram Sadak Yojana (PMGSY), NEC, BRO and Ministry of Development of North-Eastern Region (MDoNER), though some of these heads (like PMGSY) were not necessarily meant for building highways (Das 2009). Eventually, various external factors, such as delay in implementation, misuse of funds, environmental clearance, land acquisition issues and people's resistance, increased the overhead costs of these projects and created shortfalls and deficits in the central budgetary resources. The Cabinet Committee on Economic Affairs of India, a central authority, subsequently introduced a separate North Eastern States Roads Investment Program (NESRIP) in 2011 to raise capital from multiple sources through public–private partnerships. It approved an additional

Table 2.2 *Expenditure on Road Infrastructure in the Northeast under the NESRIP, 2011*

Agencies	₹ Crore	% Share in Tranche I	% Share in Tranche II
MDoNER	378.34	24.9	14.2
State governments	67.46	6.8	6.1
ADB loans	908.01	68.3	79.7
Total	1,353.81	100	100

Source: Annual Report, MDoNER, 2017–2018.

₹1,353.83 crore to meet such rising capital expenditure and created a window of opportunity to negotiate and partner with various multilateral organizations and global players. The Asian Development Bank (ADB) was the first such multilateral organization to extend financial capital towards the highway infrastructure projects of the Northeast, with a long-term objective of achieving transnational and regional economic engagements. It released funds in two tranches. Table 2.2 shows the share of capital funding of the ADB in these two tranches.

Subsequently, many other global funding agencies began to support the road and highway projects in the region with an expectation to create regional engagements. India–Japan Act East Forum, for example, was established in 2017 and promised financial assistance to these highway projects. To create potential cooperation on various connectivity and development projects in the region, Japan–India Coordination Forum (JICF) was also formed in 2017. It has identified 10 new connectivity projects in the Northeast to create cross-border mobility and tourist circuits. These projects are expected to be funded by the Japan International Cooperation Agency (JICA). Various other 'externally aided projects' (EAPs) became important windows for investment and technical support to the Northeast highway projects. For instance, the World Bank, KfW, GIZ and the Asian Infrastructure Investment Bank (AIIB) have proposed to fund about 35 EAPs (ADB-16, World Bank-8, AFD France-1, KfW-2, JICA-8) in the Northeast. About 105 more proposals (ADB-5, AIIB-2, BRICS-21, IFAD-6, JICA-57, KfW-2 and World Bank-12) are on the pipeline, and highway projects are the

Table 2.3 *Year-wise Expenditure and Number of National Highways in the Northeast*

	Number (Work in Progress)	Amount (₹ Crore)
2006–2007	100	376.56
2007–2008	98	447.11
2008–2009	111	674.7
2009–2010	NA	NA
2010–2011	77	1344
2011–2012	77	1,497
2012–2013	93	1,947.96
2013–2014	66	1,047.44
2014–2015	68	1,239.77
2015–2016	62	1,105.27
2016–2017	124	3,557.78
2017–2018	72	4,725.21

Source: Annual Report, Ministry of Road Transport and Highways, Government of India.

major components of these EAPs (NHIDCL 2018). These powerful global capital regimes are thus fast making ways into the region and proliferating with various interests, including extraction and appropriation of resources of the Northeast for gainful trade and creation of industrial corridors. Tables 2.3 and 2.4 show the expenditure patterns of the road projects in the region. Table 2.3 shows the expenditure trend of the highway projects over the last decade since the SARDP-NE was introduced, while Table 2.4 shows the state-wise and year-on-year expenditure trends of the road projects of this region.

The highways in the Northeast currently are largely coordinated and monitored by the central authority NHIDCL, with financial support from various global multilateral organizations. These partnering organizations in the long run expect to engage in various new transnational economic projects through trade and industrial corridors of the Northeast.

Table 2.4 State-wise Expenditure Trends on Road Infrastructure in the Northeast: From 2004–2005 to 2016–2017 (₹ Lakh)

	Arunachal Pradesh	Assam	Manipur	Meghalaya	Mizoram	Nagaland	Sikkim	Tripura	NER	All India
2004–2005	11,455	55,039	11,718	13,729	13,667	6,723	8,275	10,960	131,566	1,554,427
2005–2006	13,343	72,036	12,280	13,535	17,215	13,833	9,424	19,189	170,855	2,170,528
2006–2007	22,312	68,437	18,273	18,099	15,923	21,645	9,223	22,402	196,314	3,043,622
2007–2008	26,578	95,270	28,449	19,337	20,474	25,005	11,531	26,599	253,243	3,705,278
2008–2009	78,194	84,595	28,990	23,358	9,084	18,702	17,595	40,970	301,488	4,265,004
2009–2010	51,115	104,415	37,500	26,063	16,837	32,464	14,954	41,537	324,885	4,751,304
2010–2011	80,349	103,004	38,213	31,869	19,913	39,932	15,157	31,508	359,945	5,119,804
2011–2012	94,497	120,002	42,332	43,393	20,314	48,768	20,794	36,473	426,573	5,879,258
2012–2013	77,516	130,382	31,267	57,784	18,981	50,744	38,624	44,529	449,827	6,906,178
2013–2014	11,1970	166,491	37,674	63,186	19,659	47,270	38,142	52,898	537,290	8,336,126
2014–2015	13,8813	220,447	34,921	62,438	34,290	46,906	33,622	79,995	651,432	6,314,028
2015–2016 (RE)	18,8505	390,124	47,149	72,848	40,041	51,222	38,261	93,577	921,727	8,221,247
2016–2017 (BE)	11,7286	396,694	38,922	65,598	41,657	40,227	30,417	70,466	801,267	8,985,762

Source: Reserve Bank of India Publication.

ROAD INFRASTRUCTURE AND ECONOMIC GEOGRAPHY

The multilateral organizations that are investing in the highway projects of the Northeast aim to gain from India's ambitious strategy of
cross-border regional economy at both multilateral and bilateral levels.
With the new imagination and representation of the geography of the
Northeast, the vast reserve of resources of this region of India is also seen
as paramount in such larger economic integration plan. Many hydroelectric and mining projects have been identified across the highland areas
of the Northeast during the post-liberalization period for extraction and
surplus trade with the international neighbours through the newly built
highway infrastructure. These new economic prospects with circuits for
values and markets across the borders are making the geography of the
Northeast powerfully formative with an emerging territorial identity
(Bathelt and Gliickler 2003). The National Institution for Transforming
India (NITI) Aayog, the central think tank of India, has produced a
new vision in its 'Three Year Action Agenda' for the Northeast on the
areas of sustainable development, infrastructure and transit connectivity,
industry and global capital. This is an extension of the earlier central
Vision Document 2020 for the Northeast of 2008. The idea primarily is
to 'augment national wealth' and create markets for the traditional units
of the Northeast, such as sericulture, floriculture, tea plantations and silk
and handicrafts industries, by providing support through various centrally
sponsored schemes. This is to create surplus and also to enhance the aspirations of the people and integrate the local economies and the traditional
knowledge systems with the larger markets. These are the policy steps
to remove the 'economic and physical barriers' of the Northeast (NITI
Aayog 2017). However, it is being argued that such policy agenda is
likely to affect the traditional modes of production, livelihood systems
and land rights of various ethnic communities and disrupt the time-tested,
sustainable human–nature relations. This is therefore likely to generate
different societal responses and ecological ramifications through creating complexities in the changing relationship of the multiple agents of
the state, firms, investors and the people in this region (Isard 1956; King
1984; Taaffe et al. 1996).

Highway network is also seen as a symbol of state power and authority in the Northeast. It is mostly seen as part of India's defence logistics,

and it is questioned if highways actually can reduce the connectivity gap and oppression of the Northeast and provide any sustained relief to the local communities and their everyday life. As argued by Soja elsewhere, production of any new space and territory due to highway/road infrastructure needs to be essentially contextualized with the human well-being first, and then as a spatial entity in relation to the changing surroundings. The simultaneity and interwoven complexity of the social, historical and spatial dimensions of lives and their inseparability therefore need to set the goal for any new infrastructure to achieve 'spatial justice' and 'regional democracy' (Soja 2000). In the Northeast, the prospect to integrate the geography of this region with the highway network and to augment national economic gains may not rightly underpin the needs, specificities and complexities of the region. This new highway infrastructure of the Northeast is likely to co-produce new histories and spaces and bring their historical geographies to posture on the conduct of human life. It is likely to deny human well-being and 'spatial justice' and would help in re-imagining the region merely as a site for appropriation and extraction of resources and creation of multiple economic corridors to reproduce the imperial legacy with dual economies, new binaries and enclavities. This geographical criticality and the new imagination therefore would remain far behind in addressing the issues of 'underdevelopment' of the Northeast, which is largely the legacy of the post-independence Indian state. It may lead to a new category of 'lumpen-development' (Frank 1966; Stoltz and Dietz 1981) and is likely to deny any development justice to its people.

CONCLUDING OBSERVATIONS

Highway projects in the Northeast have become an important undertaking of India during the post-liberalization period, more categorically over the last 5 years, making the state's visibility and engagement 'prominent' and 'hyper'. This highway infrastructure network aims to take away the stereotypical narrative of a distant Northeast borderland. It is a complete reversal of the post-independence policy of 'non-interference', where the region was largely represented as a remote periphery in India's approach to economic nationalism. This is changing, and the Northeast is now imagined as the core geography of India, and highways are seen as the means to make it a 'connected gateway' for the country's cross-border economic engagements with the neighbouring nations and with the extended Asia–Pacific region. This new form of regionality is expected to enlarge India's scale of economic interaction and integration in the long run. It is however important to understand if this newly imagined geography can essentially override the older core–periphery debate to provide any alternative territorial identity with economic prosperity and 'spatial justice'. As multilateral regimes are the major funding agencies of these highway projects, they are expected to find new ways of partnering with the Indian state to invade this resource-rich space to extract and appropriate, and in the way reinforcing the older legacies of core and periphery.

Apart from India's economic interests, these emerging and strategic highways and bridges in the Northeast also symbolize the country's renewed thrust towards national security. This is likewise another significant reversal of the security regime of the post-independence period, when the Northeast was left as an inaccessible frontier to safeguard India from external aggression. The current impetus for highway logistics in the Northeast is a reflection of the desperation and anxiety of India to counter the dominant and forceful neighbour China. China's much-hyped infra-link project One Belt One Road has furthermore pushed India to introduce the Bharatmala Pariyojana (Project) to reach the toughest mountainous areas of Sikkim and Arunachal Pradesh in the Northeast. Highway infrastructure building is thus largely an act of power and authority with India's immense and continued geo-strategic interests in this region.

More realistically, the politics of infrastructure building on the ground is creating a new venality across the states of the region. In the highland areas, these highway and bridge networks have encroached upon people's land and resources and have interrupted their traditional livelihood practices. They have therefore created channels for many local political elites and actors to negotiate with common people and manipulate various rehabilitation processes and also siphon off central funds for compensation and rehabilitation. This is unfolding and producing a new form of highland capitalism, relegating the region to a fresh nexus of power between the state, local elites, resource regimes and the corporate global capitalists, which could be deeply embedded in injustice, oppression and marginalization of the poor and under-privileged. It is therefore potentially creating space for contestation and resistance through diluting the boundaries between varied ethnic groups. Thus, the pertinent question that needs to be raised is: how can these road projects gain the confidence of the people of the region through delivering spatial and development justice? Both historical and social relations and the diversity towards a constitutive relation of the geography and economy of the Northeast remind us that such infrastructure projects can bring optimism only if an inclusive and informed strategy is adopted. Alternatively, this may not be a costless and agreeable transformation.

REFERENCES

Anbumozhi, V., I. Kutani, and M. P. Lama. 2019. *Energising Connectivity Between Northeast India and Its Neighbours*. New Delhi: Confederation of Indian Industry.

Bathelt, H., and J. Gliickler. 2003. 'Toward a Relational Economic Geography'. *Journal of Economic Geography* 3:117–144.

Bhattacharjee, J. P., and Rakhee Bhattacharya. 2018. 'Industrial Policy in North-East India: Peripheral Realities in Post-Liberalisation Period'. *The Indian Journal of Industrial Relation* 54(2):244–257.

Bhattacharya, Rakhee. 2019. 'Introduction'. In *Developmentalism as Strategy: Interrogating Post-colonial Narratives on India's Northeast*, edited by Rakhee Bhattacharya. New Delhi: SAGE Publications. https://us.sagepub.com/en-us/nam/developmentalism-as-strategy/book269337 (accessed 8 February 2021).

Bhattacharya, R., and J. P. Bhattacharjee. 2015. 'Poverty and Inequality in India: Regional Disparities'. In *Regional Development and Public Policy Challenges in India*, edited by Rakhee Bhattacharya, 19–72. New Delhi: Springer.

Bhattacharya, Rakhee and G. Bhattacharjee. 2018. 'Poverty, Inequality and State Expenditure: Intra-Regional Disparities in North East India', in *Economic Growth in India and Its Many Dimensions*, 269-296. New Delh edited by Arup Mitai: Orient Black Swan Publishers.

Bourguignon, F., and B. Pleskovic, eds. 2008. *Re-thinking Infrastructure for Development*. Washington, DC: World Bank.

Chaube, S. K., S. Munsi, and A. Guha. 1975. 'Regional Development and the National Question in North-East India'. *Social Scientist* 4(1):40–66.

Das, Pushpita. 2009. 'Evolution of the Road Network in Northeast India: Drivers and Brakes'. *Strategic Analysis* 33(1):101–116.

Datta, S. (2012). 'The Impact of Improved Highways on Indian Firms'. *Journal of Development Economics* 99(1):46–57.

De, Prabir, S. Ghatak, and D. Kumarasamy. 2019. 'Assessing Economic Impacts of Connectivity Corridor in North East India'. *Economic and Political Weekly* 54, no. 11. https://www.epw.in/journal/2019/11/special-articles/assessing-economic-impacts-connectivity-corridors.html (accessed 8 February 2021).

Demenge, Jonathan. 2015. 'Development Theory, Regional Politics and the Unfolding of the 'Roadscpes' in Ladakh, North India'. *Journal of Infrastructure Development* 7(1):1–18.

Frank, A. G. 1966. 'The Development of Underdevelopment'. *Monthly Review* 184:17–31.

Fujita, M., P. Krugman, and A. J. Venables. 1999. *The Spatial Economy: Cities, Regions and International Trade*. Cambridge: MIT Press.

Ganguly, J. B. 2006. *An Economic History of North-East India: 1826–1947*. New Delhi: Akansha Publishing House.

Ghani, E., A. G. Goswami, and W. R. Kerr. 2016. 'Highway to Success: The Impact of the Golden Quadrilateral Project for the Location and Performance of Indian Manufacturing'. *The Economic Journal* 126(591):317–357.

Isard, W. 1956. *Location and Space-Economy*. Cambridge: MIT/John Wiley.

King, L. J. 1984. *Central Place Theory*. Beverly Hills: SAGE Publications.

National Highway and Infrastructure Development Corporation (NHIDLC). 2018. *Annual Report 2017–2018*. NHIDCL, Government of India.

Niti Aayog. 2017. *Three Years Action Agenda: 2014–15 to 2019–20*. New Delhi: Government of India.

Pachuau Joy, L. K., and Willem van Schendel. 2016. 'Borderland Histories and Northeastern India: An Introduction'. *Studies in Histories* 32(1):1–4.

Perz, Stephen G., Y. Xia, and A. Shenkin. 2014. 'Global Integration and Local Connectivity: Trans-Boundary Highway Paving and Rural-Urban Ties in the Southwestern Amazon'. *Journal of Latin American Geography* 13(3):205–239.

Rustomji, Nari. 1983. *Imperilled Frontiers: India's North-Eastern Borderlands*. New Delhi: Oxford University Press.

Scott, A. J. 1998. *Regions, and the World Economy: The Coming Shape of Global Production, Competition, and Political Order*. Oxford: Oxford University Press.

Sharma, Jayeeta. 2011. *Empire's Garden: Assam and the Making of India*. Durham: Duke University Press.

Soja, Edward W. 2000. *Postmetropolis: Critical Studies of Cities and Regions*. Oxford: Blackwell.

Stoltz Norma, Chinchilla, and J. L Dietz. 1981. 'Towards a New Understanding of Development and Underdevelopment'. *Latin American Perspective* 8(3–4):138–147.

Storper, M. 1997. *The Regional World: Territorial Development in a Global Economy*. New York: Guilford.

Taaffe, E. J., R. L. Morrill, and P. R. Gould. 1963. 'Transport Expansion in Underdeveloped Countries: A Comparative Analysis'. *Geographical Review* 53(4):503–529.

World Bank. 1994. *World Development Report 1994: Infrastructure for Development.* Washington: World Bank.

Yumnam, Jiten. 2019. 'International Financial Institutions in India's North-East: Pattern and Impact on People and Environment'. In *Developmentalism as Strategy: Interrogating Post-colonial Narratives on India's Northeast,* edited by Rakhee Bhattacharya. New Delhi: SAGE Publications. https://in.sagepub.com/en-in/sas/developmentalism-as-strategy/book269337 (accessed 8 February 2021).

Chapter 3

Irritants in India–Nepal Relations and Potential Benefits of Greater Cooperation

Bishnu Dev Pant

INTRODUCTION

Nepal and India share significant historical and cultural links. The leadership of the two countries has consistently failed to see eye to eye on issues of profound mutual importance. This chapter provides a brief historical review of the Nepal–India relationship and a summary of the irritants that have undermined this relationship. It proposes that a relationship that is based on equality and mutual respect would be beneficial to both countries. It also highlights the potential benefits of deeper economic cooperation between Nepal and Northeast India, proposing that Nepal should consider directly engaging with Northeast Indian states while continuing to work towards a resolution of its differences with New Delhi.

SHARED HISTORICAL AND CULTURAL HERITAGE

The relationship between the nations of Nepal and India underlies the shared historical and cultural heritage of its peoples. The Pashupatinath

Temple in Kathmandu is revered by Nepali and Indian Hindus alike, and the priests of this temple historically have hailed from a family of South Indian Brahmins. Similarly, while the Buddha was born in Lumbini, a town close to the Indian border in Nepal, he gave his first sermon in Sarnath, India, and died in Kushinagar, India. The Vikram Samvat, which was initiated by King Vikramaditya of India, is still used by the Government of Nepal for all official and non-official purposes. The Nepali language—once known as *khas-kura*, which translates as the speech of the *khas* hill tribesmen with whom Nepal has long been associated in South Asia's epic literature—has its base in Sanskrit, like many Indian languages. Thus, it is not surprising that Nepali has been recognized as one of the official languages of India and has been included in the Eighth Schedule to the Indian Constitution. Likewise, Hindi is spoken and understood by many people in Nepal, even though it has not yet been recognized as an official language (Upreti and Upreti 1995).

Early History

For much of its history, the region that is now Nepal was broken up into several small, independent *rajyas* or kingdoms. In the 18th century, Prithvi Narayan Shah of the house of Gorkha began a campaign to unify the *rajyas*. Shah and his descendants were able to unify much of the Himalayan region relatively quickly and began pushing into the territory of what was viewed by the British as part of their colony. Their expeditions eventually brought them into conflict with the British Raj, which was trying to establish its presence in the subcontinent at around the same time. A war between the two expanding powers began in 1814 and concluded in 1816 with the signing of the Treaty of Sugauli. The treaty ended Nepal's territorial expansion (and forced it to return conquered territories in Garhwal, Kumaon, Sikkim and Darjeeling) and set the borders of modern Nepal (Rose 1961, 201–216).

After the Treaty of Sugauli, Nepal, while remaining a sovereign and independent state, practically followed a policy of isolation and exclusion until the end of the Rana era in 1951. However, just before the Rana regime came to an end, the last Rana Prime Minister, Mohan

Shamsher J. B. Rana, signed the controversial Treaty of Peace and Friendship, 1950, with Government of India, hoping that the Indian government would support the continuation of the Rana regime.

EMERGENCE OF ANTI-INDIAN NATIONALISM

After the Rana rule in 1951, the newly formed government of Nepal endeavoured to expand Nepal's foreign policy from an India-centric model. During the isolation that followed the signing of the Treaty of Sugauli, Nepal, had become utterly dependent on India, to the extent that when Nepal exchanged goods or ideas with the world, it was only with with British India or with British India's grace. King Mahendra, who ascended to the throne in 1955, continued Nepal's 'special relationship' with India and established diplomatic relations with a number of countries. For instance, despite India's displeasure, he signed agreements with China to build the Araniko Highway, which connects Kathmandu to the Tibetan border at Kodari. In a press conference, King Mahendra proclaimed that the highway was a central plank of his economic policy, an internal matter, and therefore did not require consultation with India (Thapaliyal 2003). Because of such efforts, there was a general decline in India's leverage over Nepal. Nepal could move to a position of relative freedom in the foreign policy domain. There was a growing level of friction in Nepal's relationship with India.

In 1959, a popular government under the leadership of B. P. Koirala, who also took part in the Indian freedom struggle against British rule, was elected to Nepal's Parliament. It was expected that he could bring some improvement in India–Nepal relations. But Koirala's government was dissolved in less than 2 years' time in 1960 through a military coup by King Mahendra. King Mahendra introduced a non-party political system called the 'Panchayat System' and resumed his campaign of actively distancing Nepal from India, which continued to erode the relationship.

King Mahendra died in 1972 and was succeeded by his son, King Birendra. The young man had spent much of his formative years in the West at Eton and Harvard. But if New Delhi hoped this would give it an opening to re-establish itself in Nepal, it was in for a disappointment. In 1972, King Birendra proclaimed Nepal as a zone of peace, with the aim of keeping Nepal out of India's sphere of influence. The proposal was accepted by 100 members of the United Nations. India was not one of them. India suspected that Nepal's initiative would encourage a remodelling of the India–Nepal relationship and run against the spirit of the 1950 Treaty. India was then becoming the pre-eminent power in South Asia. It actively intervened in what was essentially a domestic Pakistani matter to facilitate the emergence of Bangladesh in 1971. In 1974, it conducted a peaceful nuclear explosion. In 1975, Sikkim was merged with the Union of India. All these distinct reminders of India's increased capability and strength forced King Birendra's resolve to protect the nation's independence against Indian influence.

In 1990, waves of protesters took to the streets to demand an end to the absolute monarchy in Nepal. The movement promised profound changes in Nepal's relationship with India. As described above, since the signing of the Treaty of Sugauli in 1950, an ideological and political connection had crystallized between nationalism, the monarchy and anti-Indianism, on the one hand, and cosmopolitanism, democracy and a softer stance vis-à-vis India, on the other. With tacit Indian support, the movement toppled the monarchy and introduced multi-party democracy to Nepal for the second time since the short-lived experiment in the 1960s.

The 1990s, thus, began with great hopes for an improved relationship between India and Nepal. There were frequent high-level visits by the leaders of the two countries. A new trade and transit treaty was signed which helped increase Nepali exports to India and reduced the trade gap between the two countries. These initiatives increased trust and reduced the hostility that had characterized the relationship in the preceding decades (Kavitha 2016). However, there continued to be friction in the relationship, and patterns of the past re-emerged.

NEW AND UNRESOLVED ISSUES

The major new and unresolved issues that have continued to complicate the Nepal–India relationship since 1990 are summarized below.

Big-Brother Attitude

Nepalis begrudge India's 'big brother' attitude towards Nepal and what they view as its meddlesome foreign policy. India holds great sway over Nepal's domestic politics and is accused of using this influence to further its vested interests. India's blockade of Nepal in 2015—shortly after the earthquake that devastated so much of the country—was an extreme exhibition of this tendency. The symbolism that Narendra Modi displayed on his first trip to Nepal in 2014 was subtler. During the visit, he reunited a Nepali migrant boy, whom he had taken care of and whose education he had sponsored in Gujarat, with his family in Nepal.[1] The gesture was warmly received in Nepal (and India), but critics asked whether it bespoke Modi and the Indian establishment's view of Nepal as a younger brother or a child desperate for Indian aid rather than an equal and sovereign nation (Soutik Biswas 2014).

The government of Nepal seems very keen to take forward the relationship with India, based on the principles of equality and mutual respect—as would apply to its relationship with China. However, being a sovereign country, Nepal would always refuse to accept India's big-brother attitude. The view of the Nepali government is that if India wants to play a role in world politics, it would have to first maintain cordial relationships with its neighbouring countries. Otherwise, its role would be questioned and challenged. Thus, there is a need for the Indian government to review its position and address the problems of that position so that the relationship between India and its neighbouring countries can be cordial.

China Factor

India views China as a geopolitical rival. The two countries are among the world's most populous countries and fastest growing economies. They have fought a war and engaged in a border skirmish as recently as 2018. They also compete for influence across Asia. Recently, India has taken advantage of Chinese missteps in Southeast Asia to gain new friends in the region. Meanwhile, China has steadily built a chain of commercial and military facilities in the Indian Ocean—dubbed as 'string of pearls'—to secure its energy supplies and contain India's strategic options during any future conflict. Thus, India has displayed a tendency to view Nepal through the lens of its policy towards China and views any deepening of ties between Nepal and China with a great deal of suspicion (Chellaney 2018).

India's position is appreciable, but only to a degree. In an age-old geopolitical rivalry, it makes sense that India would want to minimize Chinese influence in the region. But India's alarmist attitude has also alienated political actors in Nepal and reduced its ability to shape the political narrative. Indeed, indignation at India's 'big brother' attitude helped Prime Minister K. P. Oli to a landslide victory in the polls earlier in 2017. It is ultimately in India's own interests to respect Nepal's sovereign decision to have friendly relations with both its neighbours. It should also draw comfort from Nepal's profound historical and cultural bond with India—a factor that means Nepal is unlikely to become an instrument of Chinese influence in the region.

Unfair Treaties

The 1950 Indo-Nepal Treaty of Peace and Friendship was signed by the then governments of India and Nepal with a view to further strengthen the relationship between the two countries. This treaty granted Nepal preferential economic treatment and provided Nepali citizens the same economic and educational opportunities as Indian citizens in India. But this treaty has been a matter of acute controversy between the two countries (Thapaliyal 2003). Nepal's reservations are related to several clauses of the treaty, ranging from security to free movement of people across the border, particularly Article II, Article V, Article VI and Article VII. Article II requires both the countries 'to inform each other of any serious friction or misunderstanding with any neighbouring state likely to cause a breach in the friendly relations subsisting between the two governments'. Article V grants Nepal 'the right to import from or through the territory of India, arms, ammunition or warlike materials and equipment necessary for the security of Nepal'. Article VI provides for the government 'to give to the nationals of each other, in its territory, national treatment regarding participation in industrial and economic development'. Further, Article VII grants on a 'reciprocal basis, to the nationals of one country in the territories of the other the same privileges in the matter of residence, ownership of property, participation in trade and commerce etc'. Article VII, in particular, is considered by many as a threat to Nepal's interests because of the perception that it would threaten massive immigration from India and displace Nepalis from ownership of land (Upadhyaya 1995). It is unclear if this fear is borne out by the facts—and it speaks to a certain level of xenophobia in the public discourse. Nonetheless, several activists in Nepal have been pressing hard for the cancellation of the treaty for a long time.

Water

Several important rivers begin in the glacial highlands of the Himalayas and snake their way through the hills of Nepal before entering India and emptying in the Bay of Bengal or the Arabian Sea. These rivers are the source of livelihood for many communities in Nepal and North India. They have the potential to irrigate millions of hectares of land and generate electricity for meeting the increased demands for energy in both India and Nepal. Naturally, they are also a great source of tension between the two countries.

There are four treaties signed between Nepal and India regarding three large rivers. First, the Sharada River agreement was signed in 1927 between British India and Government of Nepal with respect to the Mahakali River. Second, the Koshi River agreement was signed by the governments of Nepal and India in 1954 with respect to the Koshi River. Third, the Gandak agreement was signed by the governments of Nepal and India regarding the Gandaki River in 1959. Finally, the Mahakali Treaty was signed in 1966. However, all these treaties are said to be one-sided and less favourable to Nepal; for instance, many have expressed their displeasure at the submergence of their territory and the resultant displacement caused by the Koshi Barrage. While bilateral agreements obligate India to provide compensation for flooding caused by the Koshi Barrage, there are complaints that this compensation has been inadequate. India's control and management of the barrage has also been considered as an infringement on Nepal's territorial sovereignty. Likewise, Nepal considers the Mahakali Treaty as flawed, as it lacks provisions for protecting Nepal's water rights. Nepali people feel that they are not provided with their fair share of water (Gyawali and Dixit 1999).

Security Concerns

Security has been a major concern for all successive governments of India after the British Raj. The concern stems from a fear that smugglers, terrorists and criminals can misuse India's long open border with Nepal to perpetrate their schemes and evade capture by security forces. This fear is not unfounded. Terrorists hijacked an Indian Airlines plane that departed from Kathmandu's Tribhuvan airport in December 2000. The incident raised alarm bells in the Indian security establishment about the vulnerabilities of its northern border (Thapaliyal 2003).

Despite its many benefits for economic and cultural exchange, the open border has caused problems for Nepal as well and reports of criminals and smugglers fleeing across the border are not uncommon. Separately, refugees have also taken advantage of the open border to escape inhospitable and hostile conditions elsewhere. Many Bihari Muslims from Bangladesh fled to Nepal during the Bangladesh War in 1971 via India, and many of them have remained in the country. Recently, many Rohingya Muslims from Myanmar entered Nepal via India. While Nepal shares a universal commitment to the welfare of refugees, it is strongly felt that border crossing must be scientifically regulated to balance competing demands on Nepal's limited resources—and to keep anti-social elements out.

Border Encroachment

Frequent complaints of encroachment on Nepali territory are a major source of tension between the two countries. Nepal has accused India of encroaching on key strategic areas, like Kalapani and Lipulekh (also bordering China), which lie within the Nepali territory. There is a joke that Nepali villagers who live close to the border may sleep in Nepal but awake to discover they are in India. The story is apocryphal but attests to a widespread mistrust of Indian intentions regarding the border.

Harassment of Migrant Workers

In addition, Nepali migrant workers who work in India are often harassed by the Indian border security force Sashastra Seema Bal at border
entry points. This is a long-running problem. The Indian government's
attention has been drawn to this matter several times, but the problem
persists. While there is a sizeable population of Indian migrant workers
in Nepal, there have been few documented cases of mistreatment of
such migrant workers by Nepali border officials.

Blockade

India surrounds Nepal on three sides. This in effect gives India control
over Nepal's international trade position because Nepal does not have
access to the sea. Three times in recent history, India has exploited this
strategic advantage by blockading Nepal in response to certain domestic political events in Nepal. A 5-month-long unofficial blockade in
2015 was imposed by the Indian government from October 2015 till
February 2016. The blockade took place when Nepal was still struggling to recover from the devastating earthquake of April 2015. The
blockade, which had a heavy toll on Nepal's economy not only raised
significant anti-Indian feelings in Nepal and led India to squander the
considerable goodwill Modi's outreach to Nepal had generated, but
it also contributed to bringing Nepal much closer to China. China
did not have to make much effort to neutralize the Indian influence
in Nepal. The blockade created an anti-Indian sentiment, and public
opinion towards China became more favourable (Bhattarai 2018).
Government of Nepal could not import even the most essential and
basic commodities such as petroleum products, medicine and food
products, including salt and sugar, from India or third countries, pushing Nepali people further into the poverty trap. Nepal did not have
much choice except to import these items from China. This was not
the first time that India imposed a unilateral economic blockade on
Nepal. There was a year-long blocked in 1989 and another one that
lasted a few weeks during Indira Gandhi's premiership in 1967. The
visits of Nepali Prime Minster K. P. Sharma Oli to India in April 2018

and Indian Prime Minister Narendra Modi to Nepal in May 2018 have helped heal the souring relationship to a large extent, but the suspicion and mistrust between these two countries are still visible.

REVIEW OF INDIA-NEPAL RELATIONS

In view of the increased pressure to review the Indo-Nepal relations, the governments of India and Nepal established an Eminent Persons Group (EPG) in January 2016 to revisit all bilateral agreements and submit a comprehensive report to both governments on how to reset bilateral relations. It has been felt by both countries that the 1950 Indo-Nepal Peace and Friendship Treaty requires revision, taking into account the changing international and regional scenario, including the growing concerns of the Nepalese people. To enhance people-to-people relations, Nepal and India must resolve contentious issues relating to the border, including the two major areas of dispute at Susta and Kalapani. There is also a need for construction, restoration and repair of boundary pillars and clearance of no man's land on both sides. Besides the technical aspects, the disputes at Susta and Kalapani must be resolved through dialogue by the political leadership. However, the leaders of both sides seem unwilling to take up this contentious issue. It is hoped that the EPG would be able to examine all these existing concerns and make recommendations for consideration and implementation by both the countries, so that a sound and substantive relationship can be re-established once again between the two countries.

To enhance the relationship between the two countries, India must respect Nepal's sovereignty. Mutual respect is key in bilateral relations, and India should not meddle in the internal political affairs of Nepal or panic over China's growing investment in Nepal. There are pressures on Government of India from within India to enhance ties with Nepal, based on these principles, and work with the Nepali government. Nepal must be able to address India's legitimate security concerns and assure the latter that no anti-Indian activity is conducted from within Nepal. It would need to focus on economic and development issues. With the promulgation of its new constitution in 2015 and the completion of all levels of elections in 2017, Nepal's political transition has been completed. Now the country is preparing to focus on building up its economy for advancing economic prosperity and development, for which India's support would be important.

MOVING AWAY FROM NEW DELHI- AND KATHMANDU-CENTRIC THINKING

Nepal's strategy vis-à-vis India has so far been mainly to deal with the central leadership in Delhi. As discussed above, the relationship between Kathmandu and New Delhi is complicated and has fallen short of its full potential. Thus, Nepal should contemplate furthering its relationship and outreach towards its immediate neighbours in India, including Sikkim, West Bengal, Bihar, Uttar Pradesh and Uttaranchal Pradesh. While these relationships may present their own issues, they may not be affected by the same issues undermining Nepal–India relations and could be productive for both sides.

ECONOMIC COOPERATION WITH THE NORTH EASTERN REGION

Nepal shares strong historical heritages with north-eastern states of India. This region has tremendous potential to become a gateway to Nepal for connecting with Southeast Asian and East Asian countries. One important area where north-eastern Indian states and Nepal can cooperate is tourism, as both places are renowned for their cultural heritage, diversity, natural beauty and rich natural resources. There is a growing middle class in Nepal which is interested in domestic and international tourism. A direct transportation link between Nepal and parts of Northeast India would encourage more Nepali tourists to visit Northeast India. It would also encourage tourism in the other direction. In addition, it could also facilitate religious tourism for travellers from Southeast Asia who are interested in visiting holy Buddhist sites in Nepal and North India. Thus, it is important that the governments of both India and Nepal give serious thought to this matter and consider establishing direct rail and air links between this part of India and eastern Nepal (Haokip 2010, 2011).

Northeast India is rich in spices, fruits, flowers and herbs and has considerable potential to export these products to neighbouring countries, including Nepal. Incorporating Nepal into the production chain could also lead to significant saving of costs and synergies. For this, Nepal should initiate economic diplomacy with Northeast Indian states through opening consulate offices with major trading states, so that smooth economic activities can take place with reduced non-tariff barriers and other unnecessary hindrances.

Another area where Nepal and north-eastern Indian states can cooperate is energy trade. Both the Northeast Indian states and Nepal can use their huge hydropower potential in forming a common transmission grid and exporting surplus electricity to the rest of India, Bangladesh and south-western China. However, to realize the boundless possibilities in energy cooperation between Nepal and Northeast India, crucial steps must be undertaken, the most prominent one being a political consensus at the domestic front within both countries, and at the bilateral level, at both the central and state levels. The conventional understanding that Indian interests in Nepal's hydro-power sector is guided by its strategic needs rather than energy requirements must change for the better. India can demonstrate the opposite through investing in cross-border transmission lines and reaching out to Nepal's private power producers (Rijal 2007).

GATEWAY TO SOUTHEAST ASIA

Government of India adopted the Look East policy in 1992 for enhancing partnerships with the neighbouring countries of East and Southeast Asia. China's 'One Belt, One Road' initiative and overtures by Myanmar and Thailand towards India have encouraged India to further develop its Look East policy. The policy has and will continue to boost trade and connectivity with East and Southeast Asia. A significant collateral consequence of this policy is to increase the importance of Northeast India as a gateway to Southeast Asia, particularly for the landlocked countries like Nepal and Bhutan. However, to unlock the full potential of this policy, Nepal and the Northeast Indian states must cooperate and create the institutional mechanisms to address the bottlenecks to greater connectivity, such as poor infrastructure. If realized, this vision of integration between Northeast India and Nepal would produce significant benefits for both sides, boost our collective economy and create an enormous market with a free flow of goods, services, capital and human resources—this is completely doable, with just a little political will on both sides.

CONCLUSION

Nepal and India share an ancient historical heritage, close cultural affinity, overlapping ethnic and linguistic identities, an open border permitting passport-free passage and economic linkages that have grown more significant over time. Yet, there are foundational problems in this relationship which have prevented it from fully flourishing. Resolving these problems would require India to begin viewing Nepal as an equal and sovereign nation rather than as a back door for Chinese influence, and require Nepal to adequately address India's legitimate security concerns. However, Nepal's differences with New Delhi should not obscure the potential economic benefits of greater cooperation between Nepal and the north-eastern states of India. The benefits of such cooperation have grown since the promulgation of India's Look East policy, which promises to turn Northeast India into a gateway to East and Southeast Asia's booming economies. However, securing such promise would require Nepal and India to direct institutional resources to addressing bottlenecks to greater connectivity, and particularly to improving infrastructure in the region.

NOTE

1. *The Hindu*, 31 July 2014.

REFERENCES

Bhattarai, Kamal Dev. 2018. 'Rise of China in Nepal'. *Republica*, Kathmandu, 19 July. https://myrepublica.nagariknetwork.com/news/rise-of-china-in-nepal/ (accessed 8 February 2021).

Chellaney, Brahma. 2018. 'India's Mistakes Have Allowed China to Make Inroads into Nepal'. *Hindustan Times*, New Delhi, July 6. https://www.hindustantimes.

com/analysis/india-s-mistakes-have-allowed-china-to-make-inroads-in-nepal/story-uxGLpHNMn3xKB0599aoQNI.html (accessed 8 February 2021).

Gyawali, Deepak, and Ajaya Dixt. 1999. 'Mahakali Impasse and Indo-Nepal Water Conflict'. *Economic and Political Weekly* 34(9):553–564.

Haidar, Suhasini. 'Modi meets family of Nepali foster son', *The Hindu*, 3 August 2014, https://www.thehindu.com/news/national/modi-reunites-nepalese-youth-with-parents-after-16-years/article6277127.ece

Haokip, T. 2010. 'India's Northeast Policy: Continuity and Change'. *Man and Society* 7: 86–89. https://papers.ssrn.com/sol3/papers.cfm?abstract_id=1688961 (accessed 8 May 2019).

Haokip, T. 2011. 'Essays on the Look East Policy and North-East India'. *Man and Society—A Journal of North East Studies* 8: 161–172. http://www.freewebs.com/roberthaokip/articles/essays_lep_nei.pdf (accessed 8 May 2019).

Kavitha, K. K. 2016. 'The Changing Paradigm of India–Nepal Relations: Problems and Prospects'. *Journal of Research in Business and Management* 4(5). http://www.questjournals.org/jrbm/papers/vol4-issue5/B451015.pdf (accessed 8 May 2019).

Rijal, P. 2017. 'India-Bangladesh-Myanmar Cooperation: The Gateway for Deeper regional Integration'. Integrating BIMSTEC. Symposium Conducted at Tripura, India by Ministry of External Affairs, GOI and Indian Chamber of Commerce.

Rose, Leo E. 1961. 'China and the Anglo-Nepal war: 1814–1816'. *Proceedings of the Indian History Congress* 24: 208–216. https://www.jstor.org/stable/44140753?seq=1 (accessed 8 May 2019).

Soutik Biswas. 2014. Is Indian PM Narendra Modi's 'foster son' a public relations triumph? *BBC News*, 7 August 7. https://www.bbc.com/news/world-asia-india-28673174 (accessed 8 February 2021).

Upadhyaya, S. K. 1995. 'Nepal–India Relations: The Changing Dimensions'. In *New Perspectives on India–Nepal Relations*, edited by Kali Bahadur and Mahendra P. Lama, 107–124. New Delhi: Har-Anand Publications.

Upreti, Ramakant, and B. C. Upreti. 1995. 'Changing Nature and Priorities of India's Foreign Policy vis-a-vis the India-Nepal Relations'. In *New Prospective on India Nepal Relations*, edited by Kalim Bahadur and Mahendra P. Lama, 138–162. New Delhi: Har-Anand Publication.

Thapaliyal, Sangita. 2003. *Contesting Mutual Security: Nepal–India Relations*. New Delhi: Observer Research Foundation.

Chapter 4

Indo-Bhutanese Relations
Evolution and Changing Contexts

Yedzin W. Tobgay

Many observers see Bhutan as an ancillary/feeder state of its subcontinental neighbour India. Given the perceived isolation of the Himalayan frontier states of Nepal and Bhutan, it can be said that until recent times, they were outside the ambit of the public consciousness of an international audience. This limited perception has been further compounded by President Trump's callous remarks pronouncing them as the 'Nipple' and 'Button' of India, respectively.[1] While the interaction between India and Bhutan has been one that is rich, their shared history is yet to be documented comprehensively. This chapter at the outset thus presents a comprehensive review and analysis of Indo-Bhutanese diplomatic history while also highlighting the contemporary features and issues that look at the existence of the Himalayan region as being more than one defined in relation to India. It thus also highlights the unique geography, ethnic configurations and origins of the political landscape to serve as a context to the socio-economic trajectory of relations, one that establishes the Bhutanese nation state as a complex and knowable entity in its own right and not just one that is caught between the cross hairs of the political developments of its herculean neighbours India and China.

SITUATING BHUTAN: POLITY AND ETHNIC COMPOSITION

A large and growing body of literature generally treats the country as ecologically divided into three lateral zones: (a) the southern zone, where the foothills lead to the plains of the Brahmaputra basin, which is as low as 3,000 ft above sea level; (b) the central zone, including the east and made up of valleys between the foothills and the highlands; and (c) the north-eastern zone, where the central belt works up to include the great snow-capped Himalayan ranges as high as 24,000 ft above sea level (Aris 1979, 13; Coelho 1971, 57).

If the trajectory of the 'development' of Bhutan is to be judged by its relations with its neighbours, or if the metric of growth is measured in terms of international relations in the modern sense, it is only recently that Bhutan has shed its centuries-old coat of isolation and obscurity and allowed itself a definite presence in the international arena, despite having a rich history of relations with its neighbours, often in the form of trading communities. V. H. Coelho's (1971) account of Bhutan had it 'bounded on the north by Tibet, on the west by Sikkim and the Chumbi Valley of Tibet and on the east and south by the Indian states of Assam and West Bengal'. Today, Bhutan is approximately 18,000 square miles in extent and is landlocked by China to the north across the Himalayas and by foothills that transition to the plains of India's north-eastern states to the west and south (after the annexation of Sikkim to India in 1975).

The population of Bhutan is broadly constituted by three major groups: the Ngalong of central Bhutan in the west; the Sharchops, 'Easterner', in the east; and the Lhotshampa—'Southern Borderlander' (Aris 1994, 12; Hutt 2007, 4). The Ngalong and Sharchops are generally categorized as Drukpa despite the difference in their lingua franca (Dzongkha in the west and Tsangla in the east), in contrast to the Lhotshampa, who are mostly Hindus and primarily speak in Nepali (Aris 1994, 13; Hutt 2007, 4–5).[2] Despite ambiguity in the ethnic make-up of Bhutan, the population of 750,000 has been roughly been divided into 50 per cent Drukpa, 35 per cent Lhotshampa and 15 per cent migrant tribes (Schroeder 2014, 64).

THE CREATION OF THE BHUTANESE NATION STATE: FORCES OF INTERNAL DYNAMICS AND EXTERNAL FACTORS

Perhaps Indo-Bhutanese connections can be perceived as organic since the arrival of the famous Indian Tantric practitioner/teacher Padmasambhava in the 8th century, since it is generally marked as the beginning of the early Buddhist era in Bhutanese history (Phuntsho 2013, 102). Buddhist cosmology had the Bhutanese often situating their country north of Bodh Gaya, where the Buddha attained enlightenment—the heart of our world, as Karma Phuntsho describes it. The closure of Bhutan's borders with Tibet in 1959 saw increased trade with India, and more Bhutanese pilgrims flocked to India.

The emergence of Bhutan as a unified political entity has its roots in the country's unification in the 17th century by Zhabdrung Ngawang Namgyal, a prince–abbot who was a political refugee from the Drukpa Kagyu school of Ralung in Tibet (Aris 1994, 27). Since his arrival, he is regarded to have introduced a centripetal tendency to religion and politics, combining both to establish seats of power which were considered legitimate and came to be widely accepted after extensive internecine warfare. Although Buddhism had been a dominant religion since the 8th century with the coming of Guru Padmasambhava, the adoption of the Drukpa Kagyupa school as the state religion has been attributed to reforms introduced by Zhabdrung. Thus, he is customarily regarded as founding Bhutan through assuming leadership both as supreme spiritual and as temporal head of the country in 1636. After Zhabdrung's death, state administration was handled by the tenuous leadership of an 'oligarchy of monk warriors and senior monks' who also managed the competing claims of his multiple supposed reincarnations (Sinha 1991, 106). Contestation of previously singular leadership allowed the emergence of a competing seat of power in the form of a feudal lord, the Trongsa Penlop. The latter was successful in consolidating his authority with the assistance of Bhutan's neighbour—British India.

These political developments within Bhutan existed contemporaneously with the growing power of colonial India. The British East India Company emerged as a force to be reckoned with following the

Bhutanese defeat in two Anglo-Bhutanese wars (in 1772–1773 and in 1864–1866) and the consequent signings of the Anglo-Bhutanese Treaty in 1774 and the Treaty of Sinchula in 1865 (Rahul 1980, 522). The confrontations and subsequent negotiations with the empire led to the formal demarcation of vast portions of Bhutan's southern border. Following these treaties, with the creation of a hereditary monarchy system, the Treaty of Punakha largely subsumed the earlier Sinchula treaty. Signed in 1910, the Punakha treaty guaranteed Bhutan's independence, granted the Bhutanese government an increased stipend and assumed control over Bhutanese foreign relations.

The then Trongsa Penlop, Ugyen Wangchuck, further consolidated his position as a leader and diplomat when he assisted the British in successfully securing trading routes and rights during the Younghusband mission to Lhasa (Tibet) in 1904. In a politically shrewd move by him, Ugyen Wangchuck ensured British support for dynastic rule and subsequently initiated the signing of the historic 'Genja' contract (Aris 1994, 90; Rahul 1980, 225). The signatories included representation from the monastic, political and village communities (Kinga 2009, 191–2). Such unanimous acceptance of the contract permitted the inclusion of an unambiguously hereditary character to the crown, thus perpetuating the institution of monarchy for generations to come (Aris 1994, 98). From the founding of the monarchy in 1907, the Wangchuck dynasty evidently worked towards the consolidation of monarchy and establishment of a modern state. The ruling elite sought to construct citizenship and identity to define Bhutanese sovereignty, which Karma Phuntsho (2013, 604) notes was done not out of choice but out of necessity, because of the absence of military might or economic power.

THE TUMULTUOUS 20TH CENTURY: CHANGING DYNAMICS AND THE IMPACT ON BHUTANESE POLITY

The Partition of India transformed the geopolitical context of South Asia, forcing Bhutan to reconfigure its policies and embrace the resulting changes. Its relationship with independent India remained ambiguous until the signing of the Treaty of Friendship between India and Bhutan on 8 August 1949. Many of the terms in this diplomatic treaty retained the character of the earlier Treaty of Punakha with the British. The similarity in the nature of the relationship led many onlookers to criticize India's geo-strategic planning as one that actively reinforced an element of colonial control over the region, reducing Bhutan to a colony in all but name (Ministry of External Affairs 1949; Yusufzai 2000, 314). A defining signifier of the control India exerted over Bhutan is exemplified by Article 2 of the aforementioned 'Treaty of Friendship'. The treaty is central to granting Bhutan autonomy and sovereign rights in matters of internal management but has a caveat that necessitates that its external affairs be conducted under the aegis of the Indian government. The context to this clause is that it was introduced in the Treaty of Punakha in 1910 as a temporary measure by the British to check Chinese influence over Bhutan. However, this colonial legislature persisted till 2008 and is often cited by many as a prime example of the delimitation of Bhutanese sovereignty by India (Yusufzai 2000, 314). Bhutan's debut in the modern post-II World War world of the 1950s was thus within the context of the Indo-Bhutan Treaty of Friendship and the overtures of the Communist Party of China (CCP).

The biggest geopolitical factor that must be kept in mind is the burgeoning Indo-China rivalry that backgrounds the political environment during the 1950s, which many referred to as the 'Great Game in the East' (Lintner 2015, 2). The crushing and humiliating defeat India suffered in the Sino-Indian war of 1962 shaped the foreign policies of Bhutan as well. It further increased the geopolitical significance of the countries like Bhutan and Nepal as frontier states by making relations with them a high priority on India's national security agenda. It is scant to say that the changing contexts in the modern world reshaped the role and the nature of the monarchy internally. The dominance asserted by Bhutan's gargantuan neighbours pushed the Bhutanese state to embark on a journey of self-definition that established its sociocultural environment as one that was unique and could not be conflated or assimilated with that of the cultures of the countries it was contiguous to. The processes of self-definition was one that fed into the centripetal tendencies introduced by Zhabdrung in the 17th century as it embarked on a process of 'Bhutanization'.

CHINESE INTEREST IN THE HIMALAYAN HINTERLAND AND THE LIMITS OF SOVEREIGNTY

Bhutan's historical association with the Tibetan Buddhist pantheon and its derivative culture would lead observers to assume that it would be more inclined to align with China rather than India. However, Chinese claims of Bhutan historically belonging to territories of the 'Kingdom in the Middle' threatened Bhutan's tenuous sovereignty and forced Bhutan to turn to India (Aris 1994, 99). Mao's rationale for incorporating Tibet into the People's Republic of China (PRC) was constructed along the lines of freeing an oppressed population from an entrenched feudal system. This caused a remodelling of Bhutanese subjecthood from subjects of a king under a feudal system to 'modern' citizens. Bothe (2012, 31) notes that the shift towards India's sphere of influence was also accompanied by a new narrative of modernity for state formation. This shaped the Bhutanese citizen as a bearer of 'social rights to development and political rights to representation'. It simultaneously correlated with India's strong rhetoric and commitment to democracy, which reflected the spirit of Nehruvian secularism and democracy within South Asia (Cartwright 2009, 404). It cemented the strong foundations of successful Indo-Bhutanese relations, because 'even a Bhutan politically oriented towards China would threaten the existence of India' (Patterson 1965, 385).

In contrast to the other buffer countries like Nepal, where the Shah kings diplomatically interacted with both the Indian government and the CCP, Bhutan placed its confidence exclusively in the government of India exceedingly early on after 1949. Apart from being 'advised' on matters of external and foreign policy, the Wangchuck monarchs were left to their own devices with little or no interference in internal administration. There is also no evidence of external pressure from other 'donor' states or international organizations to push for democracy either (Turner et al. 2011, 195). Thus, Bhutan was granted its own space to experiment with democratization. However, this did not mean that Bhutan was entirely immune to the effects of a changing political environment outside of its borders. The Sino–Indian War in 1962 compelled Bhutan to modernize its defence system. On the eve of the war, Bhutan's Prime Minister Jigme Dorji requested the British High Commission for 5,000 automatic rifles and 1,000 light machine guns, citing the need to 'vigorously' defend themselves should the Chinese attack Bhutan.[3] This request triggered a series of bureaucratic correspondences within the British Dominions Office.[4] Against the background of the Cold War, the British felt the need to contain the growing influence of the CCP and evinced interest in conceding to Bhutan's request for military assistance, but it was prevented because of the new Treaty of Punakha, which put Bhutan's external relations under the 'guidance' and control of the newly formed Indian government.[5]

THE GLOBAL STAGE AND REORIENTATION OF INTERNAL LEADERSHIP

Ardent alignment with India was extremely helpful for Bhutan in gaining United Nations (UN) membership in 1971.[6] New Delhi vocally endorsed Bhutan's bid for a UN seat, ensuring it a seat at the table of global politics.[7] Needless to say, this was also advantageous for India, as Bhutan also served as a reliable voting partner on UN issues. The quid pro quo character of Indo-Bhutanese relations effectively determined Bhutanese sovereignty on the international stage and reinforced the positionality of the third monarch, Jigme Dorji Wangchuck. India was the one of the first and only countries to provide development aid to Bhutan and still remains one of the largest contributors to its five-year plans (Sauvagerd 2018, 62). It seems that for the Indian government, the institution of monarchy was so deeply embedded and powerfully validated that pushing for democracy or supporting anti-royal activities would cause major destabilization. The need to deal with separatist movements in Northeast India was also a consideration for India to place its faith in the Bhutanese monarchy.

Remarkably, Jigme Dorji Wangchuck did not use the success of securing international acknowledgement of Bhutanese sovereignty to bolster his own authority. Taking note of the changes in the neighbourhood, he recognized the inevitable need for expanding democracy and embraced it. Widely regarded as the 'architect of modern Bhutan', Wangchuck initiated a series of reforms and restructuring of the existing economic and political systems to cope with the rapid changes of the modern world (Mathou 1999, 115). His agenda of decentralization and devolution continued to be carried on by his son Jigme Singye

Wangchuck, who concluded absolute monarchical rule through introducing a constitution and free parliamentary elections in 2008.

The transition from monarchy to democracy was planned from above. There was hardly any major powerful movement for democracy from the Bhutanese people. The formation of radical political parties by the Lhotshampa population demanding political changes is the only recorded episode of opposition to monarchy in contemporary Bhutanese history (Turner et al. 2011, 189). As a result, citizenship and subjecthood in Bhutan were constructed in terms of 'oneness', interpreting Bhutanese identity through a singular vision in terms of the Drukpa Kagyu Buddhist identity. The enforcement of conservative policies that promoted Bhutanese culture through the programmes such as the Driglam Namzha, a traditional code of etiquette, is also considered as one of the biggest domestic reasons that stimulated the uprising. Political changes in Nepal in 1990 and the rise of ethnic Nepali communities in Sikkim and the hill stations of West Bengal such as Darjeeling in India are also acknowledged for playing an indirect role in arousing the Lhotshampa unrest (Phuntsho 2013, 604; Duff 2015).

The role of the Indian government during this period was one of full compliance and support of the Royal Government of Bhutan. As an ally of the Bhutanese monarchy, it is notable, India denounced the democratic movement launched by the Lhotshampa dissidents. The uprising of the late 1990s by the Lhotshampa was limited in nature and swiftly terminated. But learning from the experience, the fourth monarch, Jigme Singye Wangchuck, hastened to work towards a constitutional government with the election of a new cabinet, introducing periodic votes of no confidence (Crossette 1998). The Wangchuck kings gradually advanced in creating a stable environment for democracy to thrive in, on their own terms, therefore assuring a secure place for the monarchy within such transformations.

As Stephen Cohen pointed out, India has tried to manage relations with its neighbours through a strategy of 'political transformation'—by 'making them more perfect neighbors by applauding and, at times, encouraging political reforms seen to enhance stability in the region' (Cartwright 2009, 406). Although there is no recorded evidence of a direct push for democracy by the Indian government, it is noteworthy that India's aid to Bhutan increased by 50 per cent immediately after the process of constitution making was initiated (Hutt 2006, 123). Moreover, the implementation of the Bhutanese Constitution in 2008 was accompanied by a revised friendship treaty in which Article 2, which stated that Bhutan was to be 'guided by the advice of the Government of India in regard to its external relations', was removed (Ministry of External Affairs 2007). Contrary to the Bhutanese experience, despite the several constitutions that Nepal has promulgated since the 1950s, Nepal has not managed to negotiate a revision of its Treat of Peace and Friendship to remove similar provisions like Article 2 in the Indo-Bhutanese treaty till date (Maheshwari 2018).[8]

The centralized authority and prevailing strength in the legitimacy of the Bhutanese kings made removal of monarchy an unappealing course of action despite India's commitment to democracy. At the end of the day, creation of a power vacuum and unpredictable destabilization brought forth by a weak democratic transition was not in the best interests of the Indian government, so its confidence remained solely with the Wangchuck monarchy. When Jigme Singye Wangchuck did eventually formulate an enduring political system strong enough to assume democracy and consolidate it, Indian support naturally soared.

Juxtaposed with Nepal's development debacle, Bhutan is widely considered a development success story.[9] Nepal's gross domestic product (GDP) per capita in 1960 was US$50.515 and rose to US$473.844 in 2008, whereas Bhutanese records that start in 1980 show Bhutan's GDP per capita at US$331.531 in 1980 and US$1,795.181 by 2008[10] (World Bank National Accounts Data and OECD National Accounts Data Files). In spite of the parallels in Bhutan's and Nepal's development processes, the discourse on development compares the two countries in very different registers. Earlier literatures chide the Bhutanese rulers for their cautious and slow reception to the forces of political modernization, describing them as the 'most cruel and clever lot' (Chauhan 1977, 248). Centuries of old tradition, culture and religion were blamed for chaining the Bhutanese masses in inertia and strengthening their feudal powers, but with hindsight, we can acknowledge that the Bhutanese state has successfully utilized these elements to carefully guide development and consolidate Bhutanese sovereignty.

EXPANSION OF BHUTAN'S INTERNATIONAL RELATIONS

In the 1950s, India was Bhutan's only international donor; however, Bhutan has recently begun to selectively filter viable donations from other countries over the last few decades (Crossette 1998). The first series of Bhutan's five-year plans promoted accessibility through emphasizing road construction, agriculture and healthcare, but after the mid-1970s, the state worked towards a development model based on self-reliance (Dhakhal 1987, 219). Making local districts directly responsible and accountable for progress encouraged political participation from those outside of the central bureaucratic infrastructure and worked towards increasing political efficacy. The marked shift in attitude towards development aid displays the prudent outlook of Bhutan's king and his planning commission that feared that easy access to development aid would plague Bhutan with 'Perpetual Aid Syndrome':

> Every development proposal was to be submitted through the people. The government officials would help the people to understand the feasibility and cost of specific projects. Also, it was made mandatory for the public to contribute either labor, cash, or materials. Only upon meeting these conditions can the Dzongda (district commissioner) forward plan proposals to the National Planning Commission for final approval. (Dhakhal 1987, 219)

The Bhutanese development model not only actualized decentralization away from the central administration of the government in Thimphu but also assimilated every district into the planning process. In effect, this process enhanced the political consciousness of the Bhutanese citizens and primed them for democracy, which would later ensure thorough and effective democratic consolidation. Continued emphasis on self-reliance and a sceptical view of foreign aid helped the consolidation of both monarchical and democratic administrations.

The Bhutanese government aimed to achieve the status of an autarchy by 2020. Besides Bhutan's successful development process, the country has also set itself apart through pursuing Gross National Happiness (GNH), a concept envisioned by King Jigme Singye Wangchuck in 1972 (Schroeder 2014, 107). The Bhutanese state's

reluctance to prioritize economic growth as the only metric of development led it to emphasize other multidimensional measures of evaluating
progress for four decades (Elliot 2017), defined by four main pillars: (a)
sustainable and equitable socio-economic development; (b) environmental conservation; (c) preservation and promotion of culture; and
(d) good governance. The small Himalayan country has stirred interests
on the international stage as a poster child for happiness and alternative
development (GNH Centre Bhutan).

The role of development in creating socio-economic transformation also correlates with how development aid was utilized in
the country. Bhutan's attitude towards development aid reflected
the inherent conservatism of its people, and so the country worked
towards actively reinforcing ideals of self-reliance as opposed to
dependence on foreign sources of aid.

ASSESSING THE EVOLUTION OF INDO-BHUTANESE RELATIONS: FRICTION, SALIENT FEATURES AND CRITIQUES

India and Bhutan celebrated 50 years of inaugurating diplomatic relations between the two countries in 2018 (Ura 2019, 3). The resolute
aspects of strong Indo-Bhutanese friendship are as follows: a strong
historical aversion to China; an alignment of national security agendas;
the economic tether of having their respective currencies pegged to
one another; the joint investment in hydro-power projects; and finally
a vast history of strong people-to-people connection between Indian
and Bhutanese citizens. A significant facet of the relationship between
the two countries is that of economy.

Bhutan is a member of the South Asian Free Trade Area (SAFTA)
and the Bay of Bengal Initiative for Multi-Sectoral Technical and
Economic Cooperation (BIMSTEC) and also has a free trade agreement with India (Centre for Bhutan Studies 2004). A distinguishing
characteristic of close economic and financial association is the one-
to-one peg of the Bhutanese ngultrum to the Indian rupee (INR). It
creates a stable exchange rate and is deemed as one of the most important parts of Bhutan's foreign earnings and international reserves. Such
close economic ties also subject the Bhutanese economy to the same

reverberations and consequences of Indian monetary and economic policies. This is evident in the reliance and increase of Indian imports, as India is still Bhutan's prime export and import destination and accounts for up to 85 per cent of exports and 72 per cent of imports (Lamsang 2017c). A recent irritant to this model of mutual economic benefi-cence is India's Goods and Services Tax (GST) reform, introduced on 1 July 2017. Although it is regarded as a historic tax reform expected to deliver significant economic transformation and reinforce the 'Make in India' policy, its impacts are far-reaching. Imports to Bhutan may become cheaper, benefitting Bhutanese consumers, but in retrospect, it could lead to higher imports and dominate the Bhutanese market through Indian products competing with and displacing local products. This would further exacerbate the widening trade deficit of Bhutan with India (Penjore 2019, 54–55). The adverse effects of a growing trade deficit may usher Bhutan into another 'rupee crunch crisis' like in 2012–2013, when unprecedented depletion of INR reserves led to a rationing of INR to the Bhutanese public. The ngultrum unofficially depreciated by about 10–15 per cent, but it gradually normalized within the following 2–3 years.

The most contentious issue of Indo-Bhutanese cooperation is hydro power. When India and Bhutan first embarked on developing a 'symbiotic economic relationship', the biggest pride and success was hydro-power cooperation. In the past, Bhutan and India have approved hydro-power projects before assessing the possible environmental impacts that they might have. The projects further contributed to envi-ronmental issues, such as loss of forest land, disturbance of wildlife habi-tats, multiple forms of pollution and lower crop productivity, among others. The hydro-power projects have reduced or even stopped some rivers from flowing completely. The negative environmental effects are compounded with the growing deficit in energy which Bhutan experiences. Due to environmental conditions in the winter, energy output decreases, and there is a loss of energy surplus (Premkumar 2016).[11] The result is Bhutan having to purchase energy from India. The cost of this energy is higher than that of the surplus energy that India buys from Bhutan in the summer months. The overall feasibility of hydroelectric power has come into question due to the negative economic and environmental impacts.

According to former Foreign Secretary of India Saran (2019), India has significantly gained in terms of energy security, and the delays in implementation and cost overruns of hydro-power projects stem from legitimate reasons, such as difficult terrain and design changes. He patronizes: 'as Bhutanese society becomes more prosperous, educated and aware, these projects will face greater public scrutiny. Indian entities involved in such projects will have to be more sensitive to local environmental concerns' (Saran 2019, 11). Bhutanese journalist Tenzin Lamsang has been an ardently critical voice in matters pertaining to Indo-Bhutan relations; he warns that total reliance on a single export commodity and a single buyer could lead to a growing trade deficit with India (Lamsang, Tenzing 2019). Furthermore, the new projects entail tougher financing conditions from the earlier 60 per cent grant and 40 per cent loan to the current 30 per cent grant and 70 per cent loan scheme (Lamsang 2019, 37–40). Lamsang (2017b) disclosed that India's power ministry issued Cross Border Trade of Electricity (CBTE) guidelines in 2016 which were seen to be targeting Bhutan through using India's monopoly buyer status and restricting the type of hydro-power investment that could be made in Bhutan. This severely handicaps Bhutan through the setting of future tariff rates differently from the current government-to-government scheme.

Such bilateral frictions have led to growing negative sentiments towards the Indian government. Even Saran acknowledges that some sections of the Bhutanese public have a view that their country should have a more balanced relationship with India and China. However, he clarifies that potential benefits being denied to Bhutan because of India is a 'misplaced sentiment'. During the 2018 elections, a pro-India party/candidate criterion was subtly insinuated. This was because a sudden cut of gas subsidies to Bhutan by India in 2013 is recognized as an enabling factor that led to the change in the governing party in Bhutan's second elections (Malhotra 2013). This brings us to the matter of disintegrating people-to-people relations between Indian and Bhutanese citizens.

In the spring of 2019, a video of an Indian tourist climbing and posing on a Buddhist stupa in Bhutan went viral on social media, with India facing a serious backlash (Achom 2019). This incident went on to compound already mounting apprehensions against the increasing number of regional tourists (Indian, Maldivian and Bangladeshi citizens are exempt from the USD 250 fee) visiting Bhutan. With mounting complaints against the disorderly entry and pollution caused by unsupervised movement within the country largely by Indian tourists, Bhutan's legislation is now attempting to enforce a mandatory 'sustainable development' fee of ₹1,200 per day for such regional visitors.[12]

The number of Bhutanese working and primarily studying in India has remarkably risen over the last five decades, but India is no longer the only destination open to the Bhutanese. There is a growing number of students seeking education across the region, and in Western countries, such as the United States, the United Kingdom, etc., as well. Even though India and Bhutan may be strongly aligned with one another on the diplomatic front, one cannot say that the organic people-to-people connection is the same as it used to be.

During the conflict in Dok-La between Bhutan and China in June 2017, India supported its Himalayan ally, declaring that any transgression across the Himalayas is of direct concern to Indian national security. Geospatial analysis indicates that China's foothold at Doklam Pass (more specifically, Dok-La) would risk India's access to its northeastern states through the narrow Siliguri Corridor. The disengagement of the 3-month stand-off between China and India at the Doklam Pass demonstrated the complex and strong bedrock between India and Bhutan. History shows that in the 24th border talks between Bhutan and China, a 'package deal' was proposed, where in 'return for the smaller disputed Doklam plateau, China traded a bigger territorial concession to disputed territories in central Bhutan' (Lamsang 2017a). In considering the effects of the deal on India's national security, Bhutan refused to accept such a deal. Although the Doklam border stand-off promptly concluded before the BRICS (Brazil, Russia, India, China and South Africa) Summit, Indian and Chinese troops had only withdrawn by 1,000 ft. Following the stand-off, news of the impending 25th border talks have been unavailable since 2018 (Lamsang 2009, 2019). ENODO, a risk management firm, conducted a real-time analysis of the Doklam stand-off by examining the social media platforms used by the Bhutanese citizens. The results showed '76 percent of Twitter and 65 percent of Facebook users questioned Bhutan's over-reliance on India's diplomatic channels to broker a deal with China' (Pillamarri and Subanthore 2017). Social media facilitated numerous debates over the advantages of developing relations with China; however, the change in public sentiment cannot erase from public memory the fear of Chinese annexation of Bhutan, like that of Tibet, when a wave of Tibetan refugees escaped to Bhutan in the late 1950s (Denyer 2008).

The proliferation of such negative sentiment towards India is also seen through diplomatic double standards that are adhered to by Bhutan. For example, Bhutan is the only South Asian country that is not collaborating with China in the latter's One Belt One Road (OBOR) project. Bhutan also refrained from joining the China-led Asian Infrastructure Investment Bank, even though India took a USD 1.5 billion infrastructure loan from the bank.

LOOKING TO THE FUTURE

These issues highlight a consistent policy-level approach taken by India of perceiving Bhutan as a China-adjacent entity. Like the British East India Company that feared that the Amban Chinese had the Bhutanese in their camp in the late 19th century, this is another derivative colonial legacy that India should consider abandoning (Phuntsho 2013, 530). 'There is a growing fatigue in the Bhutanese public space of Bhutan being treated like a buffer state between the two great powers of Asia, and every move made by Bhutan dissected for any possible China angle or the China Hand' (Lamsang 2019, 37–40).

Appraising the historical roots and development of Indo-Bhutanese relations allows us to review the past in retrospect, and although the Wangchuk monarchs of Bhutan were extremely provident in their alignment with the Indian government in the 1950s, this relationship is incontrovertibly disposed to evolve with the changing times and environment. India often cites its strong diplomatic relations with Bhutan as a possible template of cooperation with other South Asian countries, but it should first focus on strengthening and addressing the causes of its eroding ties.

CONCLUSION

Indo-Bhutanese connections have a rich and organic history that goes as far back as the 8th century. The colonial era largely saw Bhutanese interactions with the British manifest in the form of confrontation over the Dooars. These not only led to the signing of the treaties of Sinchula and Punakha but also initiated the creation of a strong foundation for Indo-Bhutanese diplomatic relations to thrive upon. The organic nature of an unusual yet intimate diplomatic association through close ties in the social, political and economic fronts of the two countries is now being challenged with the inevitable integration of Bhutan into the global stage during the period of globalization. India cannot ignore for long the emerging popular perception in Bhutan that 'New Delhi must accept this natural progression from pragmatism to dignity and equal treatment. Otherwise it risks unhealthy pressures building up below the calm surface' (Lamsang 2019, 40). Changing contexts would likely force both countries to reassess their agendas and change the way in which they view one another, or if old habits persist, India and Bhutan would both function according to old, realist geopolitical paradigms that have benefitted both parties for the last six decades.

NOTES

1. See *Time* (2019).
2. *Ngalong* is popularly taken to mean 'first risen', indicating the community's early adoption of Buddhism.
3. Bhutanese apprehension was caused by 2,000 Sino-Tibetan troops stationed at the northern border of Bhutan, hence explaining the 'vigorous need to defend'. See Correspondence from R. H. Belcher to Simmons 15 November 1962, DO 164/73, Dominions Office, The National Archive; Chinese Communist Troops on Northern Border of Bhutan, 2 February 1953, General CIA Records, Freedom of Information Act Electronic Reading Room.
4. These letters can be found in reference no. DO 164/73, Dominions Office, The National Archive.
5. See correspondence from C. R. Price to R. H. Belcher 5 April 1963, DO 164/73, Dominions Office, The National Archive.
6. The Kingdom of Bhutan was admitted as a member of the United Nations (UN) on 21 September 1971. This followed years of self-imposed isolation. Beginning in the early 1960s, Bhutan began a gradual process of opening itself

to the outside world through joining the Universal Postal Union in 1961 and finally becoming a member of the UN in 1971.

7. Special report in Weekly Summary for Directorate of Intelligence, 8 October 1971, General CIA Records, Freedom of Information Act Electronic Reading Room.
8. Article 2 of the Indo–Nepalese treaty states: 'the two governments have an obligation to inform each other of any serous friction or misunderstanding with any neighboring state likely to cause any breach in the friendly relations subsisting between the two governments'.
9. See The World Bank (2019).
10. The World Bank. 'GDP per Capita'. World Bank National Accounts Data and OECD National Accounts Data Files.
11. When Bhutan and India signed the 2006 umbrella agreement for hydro-power development (10 mega-projects), feasibility studies and detailed project reports were commissioned, and agreements between the governments were forged on that basis. The environmental and social impact assessments were conducted only after the signing of the agreement.

REFERENCES

Achom, Debanish. 2019. 'Indian Tourist on Camera Climbing Stupa in Bhutan Sparks Anger'. *NDTV*, New Delhi, 18 October. https://www.ndtv.com/india-news/video-shows-indian-tourist-climb-sacred-stupa-in-bhutans-dochula-detained-2119003 (accessed 8 February 2021).

Aris, Michael. 1994. *The Raven Crown: The Origins of Buddhist Monarchy in Bhutan.* London: Serindia.

Bothe, Winnie. 1992. 'The Monarch's Gift: Critical Notes on the Constitutional Process in Bhutan'. *European Bulletin of Himalayan Research* 40(2012):27–58.

Cartwright, Jan. 2009. 'India's Regional and International Support for Democracy: Rhetoric or Reality?' *Asian Survey* 49(3):403–428. doi:10.1525/as.2009.49.3.403

Centre for Bhutan Studies and IDE/JETRO. 2004. *Economic and Political Relations Between Bhutan and Neighboring Countries.* http://www.bhutanstudies.org.bt/publicationFiles/Monograph/mono-Ecnmc-Pol-Rel-Bt-Nghbrng.pdf (accessed 8 February 2021).

Chauhan, R. S. 1977. *Struggle and Change in South Asian Monarchies.* New Delhi: Chetana.

Coelho, V. H. 1971. *Sikkim and Bhutan.* New Delhi: Vikas Publications.

Crossette, Barbara. 1998. 'Bhutan Moves Toward a Constitutional Government'. *New York Times*, 12 July. https://www.nytimes.com/1998/07/12/world/bhutan-moves-toward-a-constitutional-government.html (accessed 8 February 2021).

Denyer, Simon. 2008. 'In Bhutan, Tibetan Refugees Yearn to Join Protests'. *Reuters*, 1 April. https://www.reuters.com/news/picture/in-bhutan-tibetan-refugees-yearn-to-join-idUSDEL25188520080401 (accessed 8 February 2021).

Dhakhal, D. N. S. 1987. 'Twenty-Five Years of Development in Bhutan'. *Mountain Research and Development* 7(3):219–221.

Duff, Andrew. 2015. *Sikkim: Requiem for a Himalayan Kingdom*. Edinburgh: Birlinn Limited.

Elliott, John. 2017. 'The King of Bhutan's Hopes in 1987 for Gross National Happiness'. *Riding the Elephant*. 17 December. https://ridingtheelephant.wordpress.com/2015/11/27/the-king-of-bhutans-hopes-in-1987-for-gross-national-happiness/ (accessed 8 February 2021).

Hutt, Michael. 2006. 'Nepal and Bhutan in 2005: Monarchy and Democracy, Can They Co-exist?' *Asian Survey* 46(1):120–124.

Hutt, Michael. 2007. *Unbecoming Citizens: Culture, Nationhood, and the Flight of Refugees from Bhutan*. New Delhi: Oxford India Paperbacks.

Kinga, Sonam. 2009. *Polity, Kingship, and Democracy: A Biography of the Bhutanese State*. Thimphu: Ministry of Education, Royal Government of Bhutan.

Lamsang, Tenzin. 2009. 'Bhutan-China Boundary Talks to be Held Soon'. *The Bhutanese*. 26 October. https://thebhutanese.bt/bhutan-china-boundary-talks-to-be-held-soon/ (accessed 8 February 2018).

Lamsang, Tenzin. 2017a. 'Giving Bhutan Its Due'. *The Indian Express*, 31 August. https://indianexpress.com/article/opinion/columns/giving-bhutan-its-due-doklam-standoff-india-china-relation-4821334/ (accessed 8 February 2021).

Lamsang, Tenzin. 2017b. 'India's CEA Declines Bhutan's Requests on CBTE'. *The Bhutanese*, 12 August. https://thebhutanese.bt/indias-cea-declines-bhutans-requests-on-cbte/ (accessed 8 February 2021).

Lamsang, Tenzin. 2017c. 'More than the Doklam Issue, Bhutan Worried About Hydropower Deficits.' *The Indian Express*, 30 July.

Lamsang, Tenzing, 2019. 'Bhutan-India Relations, A Bhutanese Journalist's Perspective'. In *India-Bhutan: Friendship Through the Decades and Beyond*, edited by Centre for Escalation of Peace. New Delhi: Ritinjali.

Lintner, Bertil. 2015. *Great Game East: India, China, and the Struggle for Asia's Most Volatile Frontier*. New York: Harper Collins.

Maheshwari, Dhairya. 2018. 'Nepal Seeks a Review of the Friendship Treaty with India'. *National Herald*, 11 May. https://www.nationalheraldindia.com/interview/nepal-seeks-a-review-of-the-friendship-treaty-with-india (accessed 8 February 2021).

Malhotra, Jyoti. 2013. 'India Breathes Sigh of Relief as with Elections, Bhutan Returns to the "Fold"'. *Business Standard*, 17 July. https://www.business-standard.com/article/opinion/india-breathes-sigh-of-relief-as-with-elections-bhutan-returns-to-the-fold-113071500056_1.html (accessed 8 February 2021).

Mathou, Thierry. 1999. 'Bhutan: Political Reform in a Buddhist Monarchy'. *Journal of Bhutan Studies* 19(Winter):114–145.

Ministry of External Affairs. 2007. 'India Bhutan Friendship Treaty'. 2 March. http://mea.gov.in/Images/pdf/india-bhutan-treaty-07.pdf (accessed 8 February 2021).

Patterson, George N. 1965. 'Rendez-vous in Tibet'. *Far Eastern Economic Review* 49(9):385.

Penjore, Dasho. 2019. 'Economic Life: 50 Years of Indo-Bhutan Economic Cooperation'. In *India-Bhutan: Friendship Through the Decades and Beyond*, edited by Centre for Escalation of Peace, 54–55. New Delhi: Ritinjali.

Phuntsho, Karma. 2013. *The History of Bhutan*. Gurgaon: Random House India.

Pillamarri, Akhilesh, and Aswin Subanthore. 2017. 'What Do the Bhutanese People Think About Doklam'. *The Diplomat*, 14 August. https://thediplomat.com/2017/08/what-do-the-bhutanese-people-think-about-doklam/ (accessed 8 February 2021).

Premkumar, Lakshmi. 2016. *A Study of the India-Bhutan Energy Cooperation Agreements and the Implementation of Hydropower Projects in Bhutan*. New Delhi: Vasudha Foundation.

Rahul, Ram. 1980. 'Making of Modern Bhutan'. *International Studies* 19(3):515–528.

Saran, Shyam. 2019. 'Keeping Alive a Precious Legacy, India-Bhutan Relations in the New Millennium'. In *India-Bhutan: Friendship Through the Decades and Beyond*, edited by Centre for Escalation of Peace. New Delhi: Ritinjali.

Sauvagerd, Monja. 2018. 'India's Strategies on Its Periphery: A Case Study in the India-Bhutan Relationship'. *ASIEN* 146(January):56–78.

Schroeder, Kent. 2014. 'The Politics of Gross National Happiness: Image and Practice in the Implementation of Bhutan's Multidimensional Development Strategy' (Unpublished PhD Thesis). Canada: University of Guelph.

Sinha, A. C. 1991. *Bhutan: Ethnic Identity and National Dilemma*. New Delhi: Reliance Pub House.

The World Bank. 2019. 'The World Bank in Bhutan'. 17 October. https://www.worldbank.org/en/country/bhutan/overview (accessed 8 February 2021).

Time. 2019, February 5. ' "Willful Ignorance". Inside President Trump's Troubled Intelligence Briefings'. https://time.com/5518947/donald-trump-intelligence-briefings-national-security/ (accessed 8 February 2021).

Turner, Mark, Sonam Chuki, and Jit Tshering. 2011. 'Democratization by Decree: The Case of Bhutan'. *Democratization* 18(1):184–210.

Ura, Karma. 2019. 'Introduction'. In *India-Bhutan: Friendship Through the Decades and Beyond*, edited by Centre for Escalation of Peace. New Delhi: Ritinjali.

Yusufzai, Rahimullah. 2000. 'Bhutan's Evolution and Politics of Unequal Treaties'. In *Government and Politics of Asian Countries*. New Delhi: Deep & Deep.

Chapter 5

Bangladesh and Northeast India
Planning Communication Links under GATT Article V

Abu Hena Reza Hasan and Sayada Jannatun Naim

INTRODUCTION

Before India attained independence, Northeast India was economically more developed compared to many other parts of colonial British India. This region used to produce industrial goods from the rich endowment of local natural resources, which were exported through the Chittagong port and transported by railways across territories of the Northeast and then East Bengal. However, following independence in 1947, the North Eastern Region (NER) lost communication through Chittagong harbour and Bangladesh (then East Pakistan) and became landlocked. Since then, the narrow Siliguri Corridor, known as the Chicken's Neck, has been the only way for this region to reach the rest of India. The poor connectivity of the Northeast with the rest of India and the world is a critical barrier to the economic prosperity of this region that is endowed with such natural resources as fertile land, a rich expanse of forests and substantive minerals, hydrocarbon deposits and rivers suitable for producing huge amounts of hydroelectricity.

India's Look East policy (LEP) was initiated by Prime Minister P. V. Narasimha Rao and continued by successive governments of the country to establish extensive and enduring economic and strategic relations with the nations of Southeast Asia. This initiative is an opportunity for the Northeast to become a geographical gateway for the countries in Southeast Asia and gain a connection to other countries of the world. Many analysts argue that without transit facilities through Bangladesh, the transport connectivity plans of India's LEP may not be economically efficient. Compared to that through the Siliguri Corridor, transit through Bangladesh is more economical. However, there is an acute lack of trust between Bangladesh and India regarding economic and political cooperation. Article V of the General Agreement on Tariffs and Trade (GATT)[1] may be the theoretical account of negotiations for transit facilities between Northeast India and Bangladesh to avoid lack of confidence between the two nations. The objective of this chapter is to understand the level of significance of transit through Bangladesh for India after it has built transport infrastructures within the framework of the LEP connectivity strategy, and in that context understand the relevance of Article V of the GATT for transit and connectivity negotiations between Bangladesh and India.

HISTORY OF GEOGRAPHICAL CONNECTIVITY OF THE NORTHEAST

The eight states of Northeast India share geographical borders with five countries—China, Bangladesh, Myanmar, Nepal and Bhutan. They have been connected to these bordering countries geographically since ancient times. As can be seen from maps, the geographic location of Bangladesh has almost separated the north-eastern states from the rest of India but for the narrow Chicken's Neck. Meghalaya and Assam were associated with the rest of India and other countries, mainly through Bangladesh, before India's independence. Assam had an effective transportation network to other parts of India via Bengal by river, rail and to some extent by road (Barpujari 1963; Chaudhury 2008). After independence in 1947, the north-eastern states lost communication through Bangladesh (then East Pakistan) and were disconnected from the traditional markets of produces of these states in Bangladesh, West Bengal and other parts of India (Kukreja 2016). The railway line between Dibrugarh (Assam) and Chittagong (Bangladesh) laid by British rulers had been a key transport route for the north-eastern states to export produce to the world (PWC 2014). Before independence, Tripura was connected to the rest of India through Bangladesh. Afterwards, Tripura became almost isolated from the rest of India and the other countries because of its remote location. The distance by road between Agartala and Kolkata is 1,700 km, but it was only 370 km before 1947. Besides, Tripura lost access to railway and water transportation opportunities (Directorate of Economics and Statistics 2013). Similarly, when the territories were part of British India, Naga Hills and Lushai Hills were connected to most of India through Bangladesh and the East and Southeast Asian countries through Myanmar. Prior to India's independence in 1947, the port of Chittagong in Bangladesh served as a natural gateway to many north-eastern states.

LOOK EAST POLICY OF INDIA AND ECONOMIC POTENTIALS

India's Prime Minister P. V. Narasimha Rao initiated the LEP in 1990 as a paradigm shift in the country's international economic and political relations. The LEP was planned to reduce India's dependence on the West for economic and political opportunities and develop intimate economic and trade relationships and security collaboration with the Southeast Asian countries through utilizing historic, ethnic and ideological connections (Shahin 2003). Bajpaee (2007) observed it as an opportunity for India to develop and enlarge regional markets for trade, investment and industrial expansion in an area of rapid economic growth. Table 5.1 shows that the countries within the scope of the LEP constituted 21.02 per cent of the gross domestic product (GDP) of the world in 2016, which had tripled since 2001. The region producing one-fifth of the global GDP has been achieving economic growth at a fast pace. It had 46.1 per cent and 47.6 per cent of the global population in 2016 and 2010, respectively. The relative population declined by 1.5 percentage points between 2010 and 2016. The faster GDP growth associated with the slowdown in population growth in the East and Southeast Asian countries targeted by India's LEP is an indication of the economic abundance of these countries. Prosperity increases purchasing power and expands the market. The LEP has increased the export from India to the Association of South East Asian Nations

Table 5.1 *The Gross Domestic Product of Some Asian Economies as a Percentage of Global Gross Domestic Product*

Country Name	GDP as Percentage (%) of Global GDP in Current US Dollars		
	2001	2010	2016
Bangladesh	0.16	0.17	0.29
Bhutan	0.00	0.00	0.00
China	4.02	9.25	14.77
Indonesia	0.51	1.14	1.23
India	1.44	2.51	2.98
Cambodia	0.01	0.02	0.03
Lao PDR	0.01	0.01	0.02
Myanmar	0.02	0.08	0.08
Malaysia	0.28	0.39	0.39
Nepal	0.02	0.02	0.03
Singapore	0.27	0.36	0.39
Thailand	0.36	0.52	0.54
Vietnam	0.10	0.18	0.27
Total	7.19	14.65	21.02

Source: World Development Indicators, World Bank.

(ASEAN) countries. A forecast report from Standard Chartered mentioned that Indian exports may be worth around USD 280 billion by 2024, which would be possible if India can use Myanmar as a trade corridor (Hunt 2014).

TRANSPORT INFRASTRUCTURE TO SUPPORT THE LOOK EAST POLICY

The North–South and East–West Corridor (NS-EW) is an ambitious road infrastructure project in India to connect ports, manufacturing hubs, commercial regions and cultural centres. The East–West (EW) corridor is the backbone to connect the Northeast states with the rest of India. It extends from Porbandar in Gujarat to Silchar in

Assam and has a length of 3,300 km. This EW corridor has been planned to connect the India–Myanmar–Thailand Trilateral Highway, Mekong–India Economic Corridor (MIEC), Moreh–New Delhi–Hanoi Rail Link and the Kaladan Multi-Modal Transit Transport Project with the rest of India within the framework of the LEP. The purpose of the India–Myanmar–Thailand Trilateral Highway is to boost trade and commerce among the three countries, and it subsequently would expand to Laos, Cambodia and Vietnam (Lyngdoh 2016; *Financial Express* 2017). The MIEC, planned in 2004 as part of the Bay of Bengal Initiative for Multi-Sectoral Technical and Economic Cooperation (BIMSTEC), has been modified by India as per the Mekong–Ganga Cooperation to create a network of land and sea infrastructure to promote the India–ASEAN Comprehensive Economic Partnership Agreement (CEPA) (Basu 2013). The Moreh–New Delhi–Hanoi Rail Link is planned to connect the ASEAN countries with India for trade and passenger transport using Myanmar as a gateway (Bhattacharyya and Chakraborty 2010). The Kaladan Multi-Modal Transit Transport Project seeks to connect the seaport of Kolkata with Northeast states through the Sittwe seaport in Rakhine state of Myanmar and the Kaladan river boat route (Purushothaman 2012). The refusal of Bangladesh to allow India to use the seaport of Chittagong for Northeast connectivity forced India to use Myanmar's facilities (Manipur Online 2010). It is interesting to observe that Bangladesh has opened its sea and river ports located in Chittagong, Mongla and Ashuganj to attract transits in response to the Indian connectivity plan to connect the Northeast to the other regions of the state through Sittwe Port of Myanmar (Mathur 2011).

The connectivity plans to support the LEP are converting the Northeast states into the gateway for India to ASEAN countries, using Myanmar as a land bridge. The policy of India to use Myanmar as a connectivity corridor for integrating its economy with ASEAN economies would convert Myanmar into a strategic regional logistics and trading hub (Htun, Lwin, Naing, and Tun 2011). The connectivity plan excluded Bangladesh, which had been the traditional gateway for the Northeast to other parts of India and the world, from the initiative and included Myanmar as the new gateway for the Northeast states, as shown in Figure 5.1.

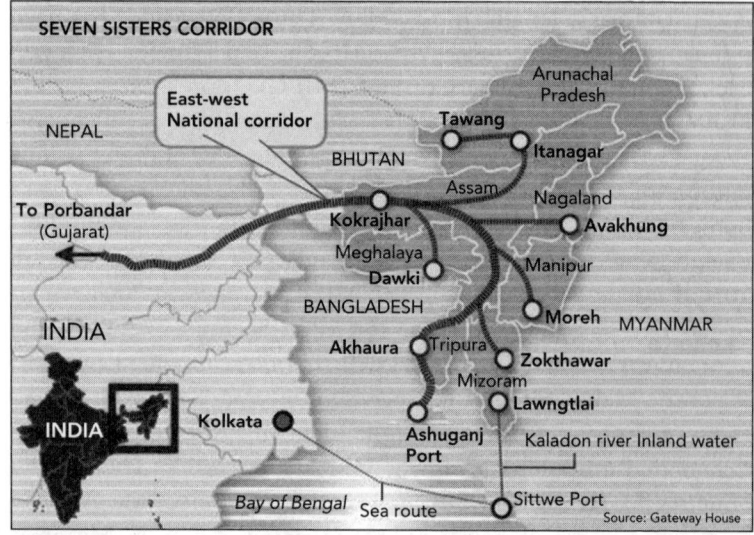

Figure 5.1 *Connectivity Plan of India within the Look East Policy Framework*

Source: SINLUNG (http://www.sinlung.com/2011/07/corridor-for-north-easts-prosperity.html).

BANGLADESH AND INDIA'S LEP: RELEVANCE OF GATT ARTICLE V

The effectiveness of the LEP excluding Bangladesh from the connectivity infrastructure has been doubted by some researchers and experts. Movement through the Siliguri Corridor may become lengthy and costly for most parts of India compared to transit/transshipment through Bangladesh. The trade between India and Southeast Asia is mostly restricted to manufactured, agricultural and primary goods. The distance between Kolkata and Agartala via Bangladesh is only 370 km, compared to 1,700 km via Siliguri. Close cooperation between India and Bangladesh for road, railway and waterway transit or transshipment may dramatically reduce transport cost and time of travel between Northeast states and other parts of India and may make the LEP more viable economically (Kathuria 2017). A survey by the Centre for Environment, Social and Policy Research (CESPR) observed that 85 per cent of the respondents in the north–eastern states had opined

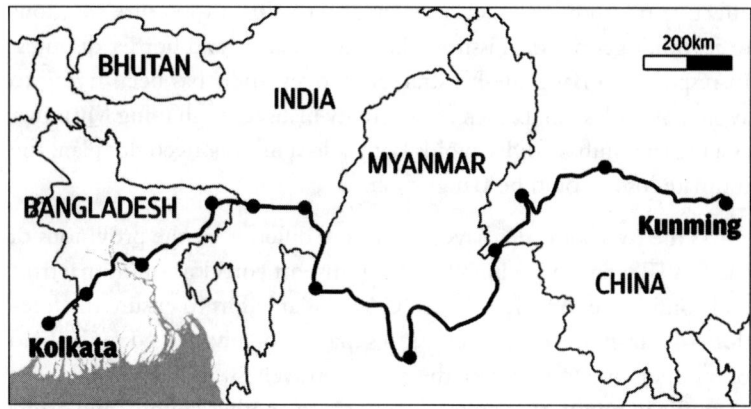

Figure 5.2 *Proposed Bangladesh–China–Indian–Myanmar Corridor Connecting India through Bangladesh, North-Eastern States and Myanmar to China*

Source: Deccan Chronicle (PTI, 2014).

that the Bangladesh–China–India–Myanmar (BCIM) corridor might be economically more viable for trade and investment (*Deccan Chronicle* 2014). Figure 5.2 shows the proposed corridor that shortens the distance between India and China significantly but needs the inclusion of Bangladesh in the LEP connectivity plans. The benefit is not one-sided. It is mutual and reciprocal. By allowing connectivity between north-eastern states and India, Bangladesh can earn foreign exchange through exporting transportation services and substantially reduce trade deficits.

The mutually beneficial relationship could not be a reality because of some unresolved issues, such as water sharing, border disputes, trade imbalance, migration of people from Bangladesh and trade protectionism adopted by India. The debate around transit versus transshipment, inadequate transport infrastructure and lack of trust between Bangladesh and India are three primary barriers for a mutually beneficial communication network (Dutta 2010). Moreover, the two countries had been negotiating with each other for a long time to ensure their economic and political interests but could not reach satisfactory conclusions because of fundamental differences in defining gains from connectivity. Bangladesh has been using its strategic location for the effective transport corridor as leverage over India to ensure solutions to its other geopolitical issues, like water sharing and border disputes. In response to Bangladesh's strategic moves, India has been trying to avoid Bangladesh in its own connectivity plans through using Myanmar as a logistics hub. If India avoids Bangladesh in a connectivity plan, the ultimate loser would be Bangladesh.

As the two countries have ideological differences, the provisions of the GATT may be a framework for transport corridors of mutual trust and confidence. Article V of the GATT is an effort to ensure the freedom of transit for trade by countries that are disadvantaged in the field of transport and to reinforce their competitiveness. It defines any goods as 'traffic in transit' that crosses a country to another country and originated from another country and which includes all functions related to movement of goods, such as transshipment, warehousing, bulk breaking and change of modes of transport. The important rules related to goods in transit are: (a) the country allowing transit cannot levy any duties on goods in transit except for reasonable charges for transportation and administrative services; (b) no country can refuse any request for transit without a valid reason; and (c) the host country for transit in goods has the right to fix the route for movement of goods, but the selected route should be convenient and economical for the country asking for transit. Under the terms of the GATT, without a user fee for services rendered to goods in transit from India, Bangladesh is not required to provide connectivity with India. India and Bangladesh should negotiate the traffic in transit within the framework of the GATT's Article V. As a member of the World Trade Organization (WTO), India has the right to request for traffic in transit, and Bangladesh has the responsibility to allow the same.

THE IMPLICATIONS OF ARTICLE V OF THE GENERAL AGREEMENT ON TARIFFS AND TRADE FOR BANGLADESH AND INDIA

India has been building transport infrastructure for connectivity with Southeast Asia through Myanmar using its Siliguri Corridor, excluding Bangladesh because of a lack of understanding and mutual trust between the two countries. Both countries understand they would gain by cooperating through reciprocal transport connectivity. India would gain competitiveness in trade and investment through additional connectivity, and Bangladesh would gain trade opportunities by using Indian transport infrastructures through Myanmar. Article V of the GATT should be the framework of negotiation between India and Bangladesh. As they are both contracting countries of WTO, Article V applies to both countries. India's Northeast, as a landlocked region, may opt for utilizing this article, and Bangladesh is obliged to negotiate accordingly. By allowing traffic in transit, Bangladesh may export transport services to India. Allowing traffic in transit for India, Bangladesh may request traffic in transit for its export and import commodities to Nepal, Bhutan, China and Southeast Asia. Southeast Asian states may be benefitted from cheaper transport facilities for moving their products to other parts of India. The NER of India and Bangladesh may become the manufacturing hubs of India for exporting goods to Southeast Asia through utilizing corridors through Bangladesh for transporting raw materials, intermediate goods and machinery from mainland India. For a win-win scenario for both countries, the GATT's Article V may be the globally accepted institutional framework for all negotiations.

TRANSIT CORRIDORS

India and Bangladesh should identify possible travel corridors for goods in transit. The identification of transit corridors should be according to Article V (2) and principles governing international transit (IRU 1998). Section 2 of Article V of GATT states: (a) the transit should be through most convenient routes in the territory of the contracting party[2]; and (b) the country cannot discriminate the transit of goods depending on place of origin, ownership of vessels and destination of the goods.

The most convenient routes must be decided after negotiation with the country requesting for transit. The routes should be practical, usable and economical to all the contracting members concerned. The means of transport and the goods carried must enjoy an identical level of access and equal conditions in the transit country. Several proposed corridors for roadways, railways and waterways across Bangladesh could be Benapole,

Banglabandha, Burimari, Sutarkandi, Ashuganj and Akhura as ports of entry and exit for goods in transit between India and Bangladesh.

TRANSIT FEES

In accordance with Article V (3), traffic originating in or destined for a territory of other contracting parties must not be subject to unnecessary delays or restrictions. In addition, contracting parties shall be exempted from customs duties and from any transit or other charges for transit, with the exception for charges for transport or those proportionate to the administrative costs incurred because of transit or to the cost of services rendered. Fixing of the transit fee is a disputed issue between Bangladesh and India, because failure to ensure economic interest may cause adverse political and social reactions. It would be the responsibility of both countries to identify the administrative and other services essential for transit and to mutually agree upon a standard procedure for charging transit fees. Bangladesh charged BDT 192 (USD 2.34) per ton of Indian traffic in transit from Kolkata to Agartala, which is considered very low by experts in Bangladesh, and the Bangladesh Tariff Commission proposed a revised transit fee of BDT 1,058 (USD 12.90) per ton (Rahman 2016). Some experts argued that congestion, road use and environmental losses must be included in the transit fee. This issue needs serious negotiations between the parties concerned.

TRADE FACILITATION

According to the WTO, trade facilitation involves simplification, modernization and harmonization of export and import processes among nations. WTO has sought to simplify customs procedures through reducing costs and making the procedures faster and more efficient. The Trade Facilitation Agreement (TFA) is a multilateral agreement that took shape in Bali's Ministerial Conference 2013 and came into effect on 22 February 2017, after two-thirds of WTO members completed their internal ratification processes. The TFA has specified the following duties for the countries: (a) simplification and harmonization of procedures for importation, exportation and required forms and documents; (b) determination of fees and charges in connection with transit; (c) preparation of list of restricted goods for transit; and (d) determination of penalty provisions for breaches of transit formalities (WTO 2014). Both Bangladesh and India have to accelerate the implementation of trade facilitation procedures while negotiating transport connectivity issues for the effective operation of goods in transit between the two countries.

TRANSIT VERSUS TRANSSHIPMENT

Both Bangladesh and India are debating the choice of transshipment versus transit. Transshipment means the consignment of goods or containers to an intermediate destination and then to another destination after changing the vehicle. Transit is the movement of goods through one country towards another destination. Article V (1) of the GATT accepts transshipment as a form of transit. There is an opinion among analysts in Bangladesh that the transport infrastructure of the country is not suitable for allowing road and rail transit to India and that transshipment may be a good alternative, as the country may export more transport services and utilize its infrastructure within its capacity (Rahman 2015). It is estimated that about USD 10 billion would be necessary to build the infrastructure for providing efficient transit to India (Ethirajan 2012). Bangladesh allowed multi-modal transshipment facilities for India to connect Kolkata with Agartala (Tripura) for transporting machinery for power plants and other goods in 2015, expecting transshipment of about 2 million tons of cargo (Mamun 2016). However, Indian transportations are reluctant to use multi-modal transshipment through Bangladesh, as they do not consider it a cost-effective option because of inadequate rail infrastructure (Mamun 2017). India has requested for cross-border container train services for traffic in transit recently (Byron 2018). Hence, transit and transshipment issues are unsettled and require more analysis and policy formulation.

CONCLUSION

Before the formation of Pakistan in 1947, Northeast India could use the territorial area of present-day Bangladesh for transport connectivity with other parts of India and many other countries. The north-eastern region and mainland of India used this geographic corridor for mutual trade and transport. However, after the independence of Bangladesh, the Northeast had to use the Siliguri Corridor to transport goods to mainland India. This time-consuming and cost-intensive transport channel affected the economic progress of the region adversely and turned it into an economically backward region. In the early 1990s, India adopted the LEP and opened up trade and transport opportunities for the north-eastern states. It started building transport infrastructure to establish transport corridors to link Southeast Asia using Myanmar as a land bridge and the Northeast as a beneficiary. The rest of the country is linked to these corridors through the congested and narrow Siliguri Corridor. Many analysts have opined that these corridors may become more efficient if India can enjoy traffic in transit through Bangladesh. However, the non-cooperating attitude of Bangladesh prompted India to avoid Bangladesh in its connectivity plan with north-eastern states through building a sea corridor between the Kolkata (India) and Sittwe (Myanmar) ports. This has created a situation in Bangladesh where it would be left out of India's connectivity plan related to the LEP. It is now necessary for Bangladesh to cooperate with India's requirements for connectivity because of the strategic initiatives of India regarding connectivity. The transit framework may be facilitated according to the GATT's Article V for freedom of transit. Both India and Bangladesh may negotiate transit corridors following Article V and explanations of it provided by WTO. They have to negotiate and formulate policies regarding transit corridors and modes of transit, transit fees, trade facilitation and choice of transit and transshipment. Rational negotiation between the two countries may result in win-win transit between them, and both may be benefitted through a scaling up of international trade and investments.

NOTES

1. This Article V of the General Agreement on Tariffs and Trade (GATT), known as the 'Freedom of Transit', is the culmination of the requirements of the Barcelona Convention and Statute on Freedom of Transit, the New York Convention on Transit Trade of Land-locked Countries, United Nations Convention on the Law of the Sea (UNCLOS) III and GATT in 1949. This article states rules related to goods in transit and movements. The rights and

limitations of home, host and destination countries are explained in this article. Details of Article V of the GATT can be read in 'The Text of the General Agreement on Tariffs and Trade' (pp. 8–9, Geneva, July 1986, https://www. wto.org/english/docs_e/legal_e/gatt47.pdf).

2. The term 'contracting party' means a member country of the World Trade Organization (WTO). The relationships among the member nations of WTO are regulated by two fundamental principles: 'general most-favoured-nation treatment' and 'national treatment'.

REFERENCES

Bajpaee, C. 2007. 'India Rediscovers East Asia'. *Asia Times*. http://www.atimes. com/atimes/South_Asia/IJ31Df01.html (accessed 15 November 2017).

Barpujari, H. K. 1963. *Assam in the Days of The Company (1826–1858)*. Guwahati: Lawyer's Book Stall.

Basu, N. 2013. 'Now, India Eyeing Mekong-India Economic Corridor'. https:// www.bilaterals.org/?now-india-eyeing-mekong-india (accessed 15 November 2017).

Bhattacharyya, A., and D. Chakraborty. 2010. 'India's Recent Infrastructure Development Initiatives: A Comparative Analysis of South and Southeast Asia'. http://dx.doi.org/10.2139/ssrn.1624466.

Byron, R. K. 2018. 'India Calls for Cross-border Container Train Service'. *The Daily Star*, Dhaka, 16 March. https://www.thedailystar.net/business/india-calls-cross-border-container-train-service-1548997 (accessed 8 February 2021).

Chaudhury, B. B. 2008. *Peasant History of Late Pre-Colonial and Colonial India*. New Delhi: Pearson Longman.

Deccan Chronicle. 2014. 'BCIM Economic Corridor Beneficial for North-East Economy' Says the Survey'. *Deccan Chronicle*, Kolkata, November 25. https://www.deccanchronicle.com/141125/nation-current-affairs/ article/%E2%80%98bcim-economic-corridor-beneficial-north-eastern-regions (accessed 8 February 2021).

Directorate of Economics and Statistics. 2013. *Economic Review of Tripura 2010–11*. Agartala: Planning (Statistics) Department, Government of Tripura.

Dutta, P. 2010. *India-Bangladesh Relations: Issues, Problems and Recent Developments*. New Delhi: Institute of Peace and Conflict Studies (IPCS).

Ethirajan, A. 2012. 'Bangladesh and India May Open Transit Networks to Boost Trade'. *BBC NEWS*. http://www.bbc.com/news/business-17229342 (accessed 16 December 2017).

Htun, K. W., N. N. Lwin, T. H. Naing, and K. Tun. 2011. 'ASEAN-India Connectivity: A Myanmar Perspective'. In *ASEAN-India Connectivity: The Comprehensive Asia Development Plan, Phase II, ERIA Research Project Report 2010–7*, edited by F. Kimura and S. Umezaki, 151–203. Jakarta: ERIA.

Hunt, L. 2014. 'Indian Trade Seen Booming With ASEAN'. *The Diplomat*, Washington DC, 26 August. https://thediplomat.com/2014/08/indian-trade-seen-booming-with-asean (accessed 8 February 2021).

IRU. 1998. 'Principles Governing International Transit According to Article V of GATT 1994'. https://www.iru.org/apps/cms-filesystem-action?file=Webnews2009/Gatt-transit.pdf (accessed 8 February 2021).

Kathuria, S. 2017. 'Boosting Business in the Bangladesh Corridor is Crucial to India's "Act East" Policy'. *Hindustan Times*, New Delhi, 26 September. https://www.hindustantimes.com/analysis/boosting-business-in-the-bangladesh-corridor-is-crucial-to-india-s-act-east-policy/story-QQil8gO-EvRPEX6XZshJteP.html#:~:text=analysis-,Boosting%20business%20in%20the%20Bangladesh%20corridor,to%20India's%20'Act%20East'%20policy&text=Broadband%20connectivity%20of%2010%20gbps,internet%20access%20in%20Northeast%20India (accessed 8 February 2021).

Kukreja, D. 2016. 'Transportation Infrastructure in the North East'. *Indian Defence Review*, New Delhi, 4 August. http://www.indiandefencereview.com /news/transportation-infrastructure-in-the-north-east/ (accessed 16 December 2017).

Lyngdoh, R. 2016. 'Highway Pact After Car Rally'. *The Telegraph India*, Kolkata, 10 August. https://www.telegraphindia.com/north-east/highway-pact-after-car-rally/cid/1417019 (accessed 8 February 2021).

Mamun, S. 2016. 'Bangladesh-India Transshipment Kicks Off.' *Dhaka Tribune*, Dhaka, 17 June. https://www.dhakatribune.com/bangladesh/2016/06/17/bangladesh-india-transshipment-kicks-off (accessed 8 February 2021).

———. 2017. 'Why are Indian Transportations Reluctant to Use Transit Through Bangladesh?' *Dhaka Tribune*, Dhaka, 28 September. https://www.dhakatribune.com/business/economy/2017/09/28/indian-transportations-reluctant-use-transit-bangladesh (accessed 8 February 2021).

Manipur Online. 2010. 'Kaladan Multi-Modal Project in Myanmar'. Imphal, December 19. http://manipuronline.com/look-east-policy/kaladan-multi-modal-project-in-myanmar/2010/12/19 (accessed 28 November 2017).

Mathur, A. 2011. 'A Corridor for Northeast's Prosperity'. *SINGLUNG NorthEast India*, 4 July. http://www.sinlung.com/2011/07/corridor-for-northeasts-prosperity.html (accessed 12 January 2018).

Purushothaman, V. 2012. 'Kaladan Multi-Modal Transit Transport Project to Link Sea Route in Myanmar with Mizoram'. *The Northeast Times*, Guwahati. https://web.archive.org/web/20120415053036/http://tntmagazine.in/news/mizoram/kaladan-multi-modal-transit-transport-project-to-link-sea-route-in-myanmar-with-mizoram/ (accessed 21 April 2021).

PWC. 2014. *Gateway to the ASEAN: India's Northeast Frontier*. New Delhi: Federation of Indian Chambers of Commerce and Industry (FICCI).

Rahman, A. 2015. 'Transhipment is Better Option than Transit'. *The Daily Star*, Dhaka, 8 March. https://www.thedailystar.net/transhipment-is-better-option-than-transit-9473 (accessed 8 February 2021).

Rahman, S. 2016. 'Transit Fee Too Low'. *The Daily Star*, Dhaka, 15 June. https://www.thedailystar.net/frontpage/transit-fee-too-low-1239754 (accessed 8 February 2021).

Shahin, S. 2003. 'India's "Look East" Policy Pays Off'. https://archive.globalpolicy.org/globaliz/econ/2003/1011indiaasean.htm (accessed 21 April 2021).

The Financial Express. 2017. 'Government, ASEAN in Talks to Take IMT Highway up to Vietnam'. *The Financial Express*, Noida, 12 December. https://www.financialexpress.com/india-news/government-asean-in-talks-to-take-imt-highway-up-to-vietnam/970887/ (accessed 8 February 2021).

WTO. 2014. 'Protocol Amending the Marrakesh Agreement Establishing the World Trade Organization: Agreement on Trade Facilitation (WT/L/940)'. https://www.wto.org/english/tratop_e/tradfa_e/tradfa_e.htm (accessed 12 January 2018).

Chapter 6

Yunnan and Northeast India
Chinese Perspective on Sub-regional Development

Hu Xiaowen

INTRODUCTION

In the era of de-globalization, sub-regional economic cooperation has gradually become the new engine of regional development. Under sub-regional cooperation, peripheral border areas take the leading roles in strengthening economic and trade cooperation between countries. Yunnan in China and the North Eastern Region (NER) in India are landlocked regions. For decades, they have remained underdeveloped and backward. However, in recent years, both China and India have taken new policy initiatives that might change the status of their land-locked regions.

In 1999, the Chinese government proposed the strategy of developing its western region. Over the past years, the western provinces have achieved significant results in infrastructure and ecological-environment construction (Wang and Wei 2003). In the 1990s, China launched the 'border opening up policy', aimed at developing peripheral and landlocked provinces. Yunnan seized the opportunity and started various regional cooperation projects: the Greater Mekong Subregion (GMS) cooperation (1992), the 'golden quadrilateral' sub-regional cooperation (1993), the Mekong River

Basin cooperation for sustainable development (1995), the ASEAN Mekong Basin Development Cooperation (1996), the Bangladesh–China–India–Myanmar Economic Corridor (BCIM) (1999), the China (Yunnan)–northern Thailand cooperation mechanism (2004), the Yunnan–northern Laos cooperation working group mechanism (2004), the China–South Asia Business Forum (2004), the Burma cooperation business forum (2007) and the China, Burma and Vietnam Cross-border Economic Cooperation Zones (2008). In 2009, when President Hu Jintao visited Yunnan, he proposed the gateway strategy for the first time, saying Yunnan needed to deepen and extend itself through opening up to the surrounding areas, take advantage of its strategic location linking it with South Asia and Southeast Asia and be the gateway to its southwest (Li 2011).

Likewise, in India, the government launched the 'Look East Policy' (LEP) in the early 1990s, which initially aimed at enhancing economic ties with Southeast Asian countries. After the second phase of the LEP, India has expanded the policy to include non-ASEAN (Association of South East Asian Nations) countries in East and Southeast Asia. In the early 2000s, the government declared development of the NER as one objective of the LEP agenda. NER Vision 2020, which pleaded for 'making the Look East Policy meaningful' for the region, emphasized linking the NER with ASEAN by sea routes through Bangladesh and 'land routes through Myanmar and China' (Downie 2014). It foresees the possibility of connecting India with China through the NER and West Bengal. The proposed BCIM has the potential to promote the interest of development of the landlocked regions. Against the background of the BCIM initiative, this chapter examines the developmental initiatives in China's Yunnan and India's NER and explores the possibility of better and beneficial interactions between these two landlocked regions.

SUB-REGIONALISM AND REGIONAL COOPERATION

Sub-regional cooperation studies emerged in the late 1980s after the Cold War. Most countries started paying attention to economic development, trade liberalization, production internationalization and financial integration. Voices seeking Asia–Pacific economic integration and other kinds of regional cooperation started gaining currency because of diverse economic and political backgrounds. Economic integration in Asia is not as easy as that in Europe. Attention was drawn to sub-regional cooperation when the Asian Development Bank (ADB) started emphasizing it. The past two decades witnessed various sub-regional cooperation mechanisms, such as the Indonesia–Malaysia–Singapore Growth Triangle (IMS-GT) that started off in 1989, the GMS cooperation launched in 1992 and the South Asia Growth Quadrangle (SAGQ) started in 1997.

Sub-regionalism requires allocation and use of resources by specific geographical units within several nation states. Borders are closed in the traditional sense. With the overall easing of international political conflicts, both sides of borders began to set up open ports at the borders to meet the needs of the residents of the border areas. Therefore, the role of the border has changed from 'barrier' to 'intermediary'. The economic function of border areas has been activated. To promote cross-border development, the state gives unprecedented attention to sub-regional cooperation. For the sake of stimulating local economic development, the local governments also wish to promote cross-border cooperation (Liu 2014).

Under sub-regional cooperation, border areas take the crucial role of carrying out economic and trade cooperation between countries. With policy coordination between countries, resources can be allocated beyond borders, and therefore, the overall competitiveness of sub-regions can be enhanced and peripheral areas can be transformed into core areas. In this context, it is necessary to explore the role of different actors and the mechanisms in place (Liu 2014).

This chapter tries to assess the prospects of sub-regional cooperation between Yunnan and BCIM-related areas in India and, taking the case of Yunnan and three states (Assam, Manipur and West Bengal) in India, analyse the role of local governments and the existing policy initiatives. It examines whether Yunnan and BCIM-related areas in India have the willingness, capabilities and policy support to develop as export-oriented economies.

BASIC CONDITIONS NECESSARY FOR AN EXPORT-ORIENTED ECONOMY

Industrial Structure

After India's independence, the development gap between the Northeast and other regions gradually widened. Economic and social development in Northeast India is relatively low. Agriculture accounts for a large proportion of the gross state domestic product (GSDP). The NER constitutes 9 per cent of India's geographical area but contributes only 3 per cent to the country's gross domestic product (GDP). Between 1999 and 2009, the average annual growth rate of GDP in India was 7.2 per cent, while its corresponding figure for Northeast India was only 6.67 per cent.[1]

Take, for example, the three major states of Assam, Manipur and West Bengal that are involved in the BCIM. The total GSDP of these three states was about US$119.6 billion by 2012, accounting for 6.5 per cent of the country's GDP. The average per capita of GSDP of these three states was USD 941, about 63 per cent of the national average. In terms of the industrial structure, the ratio of the three major industries (agriculture, industry, service trade) in these three states was 27.96:13.99:58.03, while the corresponding ratio for India was 17.39:17.64:64.97.[2] It can be seen that the proportion of agriculture in these three states is higher than that in the entire country, while the proportions of industry and service trade are lower than those of the entire country.

Trade System

In terms of economic types, the total import and export volume of Assam, Manipur and West Bengal in 2012 was only USD 580 million. Their foreign direct investments amounted to only about USD 31.8 million, and their foreign trade accounted for only 4.8 per cent of the GDP. The proportion of foreign direct investments in the GDP was only 0.26 per cent. The proportion of foreign trade and foreign direct investment in the three states' economy is very small. Therefore, it can be said that the three states of India are typical inward-oriented economies.

However, the three states above have a sizeable cumulative population of about 120 million. The enormous population means a rich labour force and a huge potential market. However, subject to historical and policy restraints, the development of the three states is relatively low. If the Indian government can adjust the relevant internal and external policies, the development potential of the three states can be explored rapidly.

Infrastructure

In the NER, lack of the facilities such as storage, warehousing and transportation affects mobilization of goods (Raimedhi 2015). The total length of the road network in the Northeast of about 377,000 km contributes about 9.94 per cent of the total roads in the country. In terms of the quality of the roads in the region, merely 29 per cent of the total roads are surfaced. The Northeast's international borders stretch 5,182 km. Internally, the low road density is exacerbated by the shoddy quality of the existing roads, while railway services only reach three stations in the entire region (Downie 2014).

To improve the traffic situation in the NER, the Indian government has made several efforts over the past years. The National Highways Development Project (NHDP) has come out with the Special Accelerated Road Development Programme in the North East (SARDP-NE) for the development/improvement of over 10,000 km of roads in the north-eastern states. The Ministry of Road Transport and Highways (MoRTH) has been paying special attention to the development of national highways in the region and has allocated 10 per cent of the total funds for the NER (Nandy 2014). Besides, the central government has launched a scheme in the last decade, the Pradhan Mantri Gram Sadak Yojana (PMGSY), to invest in the NER's roadways sector. The scheme aims to provide road connectivity to all villages with a population of over 500 persons in the plains and those with a population of over 250 persons in hilly areas (Nandy 2014).

While the projects cited above have made substantial progress, some proposed projects are still in progress, and some only on paper. The revival of Ledo Road (Stilwell Road), connecting Ledo, the Kaladan Multi-Modal Transit Transport Project and the Trans-Asian Railway, once completed, could open up a new window for development of the landlocked Northeast states of India. Various regional initiatives, such as the BCIM, the Bay of Bengal Initiative for Multi-Sectoral Technical and Economic Cooperation (BIMSTEC) and the India–Myanmar–Thailand Trilateral Highway (IMTTH), projects that aim to link the markets of South and Southeast Asia, are in initial stages of execution.

The total railway network in the NER is of length 2,602 km (as on 2011), which is only about 4 per cent of the total rail network of the country. In June 2014, the World Bank provided US$107 million in interest-free loans to Mizoram of India for road construction, aiming at facilitating transportation and connectivity between the states of Northeast India and Myanmar and Bangladesh. It is reported that besides the India–Myanmar border road, the planned road expansion involves another road from Northeast India to Myanmar and Bangladesh (Nandy 2014). On 26 March 2015, Myanmar's Vice Minister of Construction Wu Su Ding revealed that the Indian government intends to provide Myanmar with USD 257 million in help to build and upgrade the 402-mile border road, which would be helpful for developing the India–Myanmar trade channel.[3]

Two projects already underway are the Kaladan Multi-Modal Transit Transport Project and the IMTTH. The Kaladan Multi-Modal Transit Transport Project links Mizoram to Myanmar's port of Sittwe, which India is helping develop (Nandy 2014). The project is being financed by the Indian government under the category of 'Aid to Myanmar'. The IMTTH was conceived at an India–Myanmar–Thailand Trilateral Ministerial Meeting in Yangon in 2002. The project proposes construction of a 1,360-km highway from Moreh and Tamu on the India–Myanmar border to Mae Sot on the Myanmar–Thailand border (Iyer 2017). The IMTTH project is considered as an example of triangular road diplomacy between India, Myanmar and Thailand. The deadline for this project has been set as 2020.

As far as connectivity with Bangladesh is concerned, India has built a bridge over the Feni River in Tripura to ferry heavy machines and goods to and from the north-eastern states and the rest of India via Bangladesh through the Chittagong international port.[4] Ingress to Chittagong port and opening up of the inland water route could lead to an economic resurgence of the region. Many experts feel that apart from creating transport and infrastructure, there is a need to strengthen the institutional support for enhancing the existing level of trade and economic linkages in the NER (DONER et al. 2012).

Compared with other states in eastern and north-eastern India, West Bengal has better traffic conditions. The road density in West Bengal is higher than the national average in India. As of 2011, the highway mileage of West Bengal was 92,023 km, including 2,578 km of national roads, 2,393 km of state roads and 4,481 km of railways. Kolkata is the regional headquarters of India's Eastern Railway and South Eastern Railway. It has three main international airports, and its principal port in West Bengal state, up the Ganges River, leads directly to Assam.

Support to Policies and Projects

Government of India has started many measures to promote economic and social development in Northeast India. In 1971, the North Eastern Council (NEC) was created to advise the central government and start development projects for the balanced growth of the NER. It also coordinates development plans between the states. In addition, Government of India accorded 'special state status' to all Northeast states and has formulated preferential policies for them in terms of financial support. Since 1998, all central government ministries have been earmarking 10 per cent of their annual budgets for north-eastern states.[5]

The Ministry of Development of North Eastern Region (MDoNER) was established in September 2001 to act as the nodal department of the central government to deal with matters related to the socio-economic development of the eight states of Northeast India. According to MDoNER, between 1998 and 2006, more than ₹42,600 crore was allocated for the NER (Varma 2009). Similarly, the North Eastern Development Finance Corporation Ltd (NEDFi) was established in 2002 to provide financial assistance to micro, small, medium and large enterprises for setting up industrial, infrastructure and agri-allied projects in the NER of India and also for microfinance through microfinance institutions or non-governmental organizations (NGOs).[6]

As part of the country's LEP, for the last 30 years, policymakers in India have been making efforts to strengthen ties with the country's eastern neighbours. Development of the Northeast has been one of the stated objectives of the LEP. NER Vision 2020, a landmark policy planning document released in 2008, declared that for 'making the Look East Policy meaningful', the region should have access to ASEAN, by sea routes through Bangladesh and 'land routes through Myanmar and China' (Downie 2014). In January 2000, the Indian prime minister's announcement of the development of exports from the north-eastern region led to the setting up of an Export Development Fund (EDF), the primary objective of which was promoting exports from the region. Forty-seven projects have been sanctioned under the EDF so far. Passion fruit cultivation in Mizoram and Nagaland, *safed musli* cultivation in Assam, ginger cultivation in Manipur and Nagaland, organic farms in Nagaland and Tripura, etc., are some projects approved by the EDF. Additionally, Agri Export Zones (AEZs) have been set up in Tripura for pineapples, in Sikkim for floriculture, orchids, ginger and cherry and in Assam for fresh and processed ginger (Raimedhi 2015).

The development of north-eastern India has not been linked to Chinese investment. China's president Xi Jinping's 2014 visit to India brought US$20 billion in Chinese investment, which was a marked increase over the cumulative US$400 million invested over the past 14 years, but all of it was earmarked for Gujarat and Maharashtra. Referring to it, Assam's then Congress Chief Minister Tarun Gogoi said, 'We do not have any problem with investment in Gujarat, but why not Assam? We should also get a share of China's investment in India' (Downie 2014). A survey by the Assam-based Centre for Environment, Social, and Policy Research in 2014 shows broad enthusiasm for the corridor among politicians, journalists, officials and academics in the north-eastern states. In November 2019, at the Kolkata to Kunming Forum, Indian Consul-General in Guangzhou K. Nagaraj Naidu blamed the north-eastern states for failing to press their claims effectively. Northeast India 'needs to take ownership of improving connectivity', he said, adding that these states 'lack a concerted strategy for how to engage with China' (Downie 2014).

In contrast, Yunnan province and Kolkata have begun to establish a closer relationship. Kunming has direct flights to Kolkata, and the two agreed to establish sister-city status during former Prime Minister Manmohan Singh's visit to China in October 2013. China has been particularly interested in strengthening links with West Bengal. Thirty Chinese companies were present at the Bengal Global Business Summit held on 16–17 January 2018. On one occasion, China has also invited Mamata Banerjee, West Bengal's chief minister, over to visit China. As one Indian scholar suggested, Indian government should encourage interactions between eastern Indian states and Chinese provinces. Interactions between Indian states and Chinese provinces are likely to lead to a win-win situation for both sides (Maini 2018).

BASIC CONDITIONS THAT TURNED YUNNAN INTO AN EXPORT-ORIENTED ECONOMY

Industrial Structure

The economic development of Yunnan has shifted from agriculture and industry to service sector. In Yunnan province, the primary sector's contribution to the GDP declined from 37.22 per cent in 1990 to 15.53 per cent in 2014; the proportion of that of the secondary sector increased to 41.21 per cent by the end of 2014—about half of Yunnan's GDP. However, the proportion of contribution of the industrial sector is lower than the national level by nearly 5 percentage points, indicating that Yunnan is still in the middle stage of industrialization. It can be said that after more than 30 years of development, the industrial and service sectors have developed steadily in Yunnan.

After implementing the Western Development Strategy, Yunnan's government realized the necessity to develop economies with their own characteristics and helpful industries. The secretary of the Yunnan

Provincial Party Committee, in his latest article published in *Qiushi*, declared that Yunnan's comprehensive strength had increased rapidly and that its economic growth ranked at the forefront of the country. Still, he insisted that Yunnan should work hard to ensure industrial optimization and upgradation, promote the transformation and upgradation of the traditional industries such as tobacco, tourism, energy, biotechnology, non-ferrous metals, etc., expand production of bio-medicines and health products, electronics, advanced equipment, food and consumer goods and promote green energy, green food and healthy living destinations (Hao 2020).

Trade System

At present, international trade in Yunnan includes general trade, processing and assembling trade and a small border trade, which are all mutually complementary to one another. Yunnan realizes that growth of trade helps in domestic and international industrial division of labour. In March 2017, the Yunnan government released a plan document titled 'Implementation Plan on Building Yunnan as an Economic and Trade Center towards the South and Southeast Asia'. The plan aims at promoting connectivity with the surrounding areas and developing export-oriented economic development zones. It encourages Yunnan's enterprises to invest in South Asian and Southeast Asian countries and integrate into global supply and value chains.[7]

Infrastructure

Speeding up infrastructure construction is essential for developing western region. Over the past 10 years, the central government has sped up the development of infrastructure in the western region and addressed the issues of transportation, water conservancy, energy and municipal administration through enhancing fiscal expenditure and allocating special funds. Between 2000 and 2012, under the Western Development Strategy, 187 projects valued at 3.68 trillion yuan were started.[8] The proportion of investment in the western region to the national investment is increasing year by year; in 2011, the proportion was 23.1 per cent, 3.5 percentage points higher than in 2002.

In recent years, infrastructure construction in Yunnan has made considerable progress. At the end of the 'Twelfth Five-Year Plan', the road length in Yunnan reached 236,000 km, of which high-grade highways (secondary roads and above) accounted for 16,200 km and expressways 4,005 km (Transport Department of Yunnan Government 2019). The length of railways was nearly 3,000 km. The Shanghai–Kunming high-speed railway has become the longest high-speed railway line in China. The civil aviation industry in Yunnan has developed rapidly. There are 13 airports in Yunnan now, and even the remote cities have airlines connected with Kunming. Some famous travel destinations, such as Shangri-La, Xishuangbanna and Tengchong, are connected by flights.

At present, the transport facilities are gradually improving. With Kunming as a destination, efforts are on to develop a three-dimensional transportation network of highways, railways and aviation. The cable transmission system linking Yunnan, Myanmar and Laos has been completed. The China–Myanmar oil pipeline has been built, and the port facilities have been considerably improved. Seventeen road ports have been opened linking with neighbouring countries, and 24 international passenger and cargo transportation lines have been opened. From all these initiatives, it is evident that Yunnan is emerging as an export-oriented economy.

Policy Support

The report of the 19th National Congress of the Communist Party of China in 2017 proposed that China would deepen its relations with neighbouring countries under the concept of building good neighbourly relationships and partnerships. China supports the multilateral trading system, promotes the construction of free-trade zones and seeks to build an open global economy (CPC 2019). In addition, the Chinese government has also issued a series of policy directions to sustain economic growth and to reform local government. The central government continues to invest in livelihood projects, infrastructure and ecological environment in the western region (The State Council 2019). All the above policies and supports will play an active role in promoting the transportation infrastructure and deepening regional cooperation between Yunnan and its neighbouring countries.

In 2015, President Xi Jinping visited Yunnan and positioned it as China's gateway to South and Southeast Asia. He also emphasized Yunnan's role in the 'Belt and Road Initiative' (BRI) (People News 2015). In 2012, the China–South Asia Expo was organized in Kunming. Once the BRI materializes, Yunnan is expected to emerge as a frontline region in South and Southeast Asia.

POSSIBILITIES AND FEASIBILITIES OF COOPERATION BETWEEN YUNNAN AND THE NORTH EASTERN REGION

Yunnan and the NER are geographically close to each other. Both are peripheral and economically and socially backward regions in their respective countries. They occupy a relatively marginal position in their country's economies. However, in recent years, both Yunnan and India's NER have been making efforts to develop into export-oriented economies.

Over the last decades, the trade volume between China and India has experienced rapid growth. However, much of the increase in trade volume is contributed by economically developed provinces and states. Bilateral trade between the border or peripheral areas, especially between China's south-western provinces and India's north-eastern states, is negligible. The Act East policy of India and BRI of China aim at rescuing these landlocked backward economy states and helping them overcome their landlocked situation and backward roles and emerge as new growth engines in the region. Considering these basics, one can assume the possibilities of cooperation between Yunnan and the NER of India in the following areas.

INDUSTRIAL COOPERATION

Both Yunnan and India's NER are involved in trade of primary products. From the perspective of the commodity structure of the trade between Yunnan and India, the trade commodities between the two sides are mainly primary products, but it can be seen from these commodities that Yunnan and India are highly complementary in resource products, which indicates a better prospect for trade development. Yunnan exports phosphate rock and phosphate chemical products to India, while India mainly exports iron ore and other mineral products to Yunnan, and both sides give full play to their comparative advantages.

Even at the regional level, similar problems exist between Yunnan and BCIM-related areas. Yunnan belongs to an economically underdeveloped region; its economic development is in transition from the early phase to the middle phase of industrialization. Most states in the BCIM-related areas in India, barring West Bengal that is outside the NER, are still in the early stage of industrialization; the primary sector accounts for a large proportion of their GSDP. Hence, the development experience of Yunnan can be used as a reference while initiating policies for development of the areas covered by BCIM in India, especially in Northeast India. The technologies and ideas for developing characteristic agriculture, energy and minerals in Yunnan can be applied for development of the NER. Both Yunnan and the NER may undertake cooperative research for mutual development (Lu 2019).

There is abundant scope for cooperation between Yunnan and India's NER. Development of a sustainable resource processing industry is an effective way to change the underdeveloped economy of the Northeast. Yunnan province has been doing a lot of work over a long time and has accumulated rich experience in the development of resources. It may cooperate with the NER in forest processing, cultivation of ornamental flowers, processing of forest and agricultural products, production of medicines and development of minerals (Lv 2009).

Regarding exchange of development experience and development model, Chinese experts pointed out that the academic circles of the two sides can carry out exchanges and cooperation in the fields of economics, tourism, sociology, literature and art, history and pedagogy and can conduct theoretical discussions on economic development and social development in border areas. India may take cognizance of the development experience of Yunnan province and come out with a development road map and model suited to the reality of BCIM-related areas, especially Northeast India (Lv 2009). As one Indian scholar put it, 'Yunnan is a mirror of India's Northeast, The Indian Northeast's competitive advantages lie precisely in the same areas as Yunnan's: tourism and as a gateway to the East. Yet these advantages remain unexploited' (Aiyar 2016).

COOPERATION ON INFRASTRUCTURE AND CONNECTIVITY

By promoting trade through land routes, the landlocked Yunnan and India's NER could overcome underdevelopment. But unfortunately, much of the trade between India and China takes place through sea routes. However, of late, China has been insisting on land connectivity with neighbouring countries and has achieved positive results. China's YuXinOu Railway route[9] connecting it to Germany via Central Europe shortens the distance and travel time by more than half. Even in terms of cost, the land transport fare is found to be less than the sea transportation fare (Che 2010). The BCIM countries can learn from the experience of the YuXinOu model.

The BCIM proposes three routes connecting the four countries (Bangladesh, China, India, Myanmar): (a) the north route that proposes reviving the China–India road, connecting Kunming, Baoshan, Tengchong and Houqiao (China–Myanmar border), Myitkyina (Myanmar) and Ledo (India); (b) the middle route of Kunming–Ruili, connecting Yunnan's border city with Bhamo, Lashio, Mandalay and Tamu (Myanmar), Imphal (India), Sylhet and Dhaka (Bangladesh) and Kolkata (India); and (c) the south route linking Kunming and Ruili (Yunnan's border city with Myanmar) with Lashio, Mandalay, Meiktila and Magway (Myanmar), Chittagong and Dhaka (Bangladesh) and finally Kolkata (India) (Rahmatullah 2011). To popularize the routes, the four countries have organized two rounds of a car rally since 2013. On its part, China has taken positive steps in pursuit of the BCIM project. It has completed construction of roads and railways in domestic sections and invested in some construction projects abroad. The BCIM initiative has received considerable support among politicians, journalists, officials and academics (Downie 2014). Efforts are on to connect the states and major cities in the NER; however, the progress has been slow. In contrast, China is keen on the development of Yunnan. As Jin Cheng, director general of the province's International Regional

Cooperation Office, put it, 'We want to full use our region's cultural and ethnic diversity to develop tourism and we want to make Yunnan China's gateway to South Asia and South East Asia'.

On its part, China looks at the proposed BCIM routes as the revival of the traditional Silk Route. For over 2,000 years, Yunnan had been China's gateway to Southeast Asia and South Asia. The traditional Southern Silk Road began from Sichuan, passed through Yunnan to Myitkyina (Burma) and extended to India. During the Tang dynasty, merchants used to travel between Yunnan and India through two roads: one passed through Chengdu, Dali, Longling, Ruili and Shan State (Myanmar) and Manipur (India) (Shengda 1993), and the other road passed through Chengdu, Dali, Baoshan, Tengchong, Myitkyina and Mogaung (Myanmar) and Assam (India). The Silk Route, which merchants belonging to different ethnic communities used to do their business, also helped in promoting economic and cultural exchanges between Yunnan and the NER of India (Gongbu 2018). During the Second World War, the Stilwell Road, the Hump and the Sino-Indian oil pipeline bridged the material supply chain between South Asia and China and played a significant role in the anti-fascist war against the Japanese invaders in the Chinese battlefield. Remembering these common historical and cultural experiences would go a long way in strengthening the bonds between Yunnan and India's NER.

CONCLUSION

Among other objectives, the LEP/Act East policy initiated by India and the BCIM proposed by China have the development of backward regions as one of their goals. India and China resumed normal trade relations in the 1980s. Over the last decades, the trade volume between China and India has experienced rapid growth. However, much of the contribution to the trade volume is made by economically developed provinces and states. Bilateral trade between China's south-western provinces and BCIM areas in India's NER is negligible. Under the 'Act East policy' of India and BCIM, these traditional backward economies are expected to overcome their landlocked situation and become new growth engines of the region.

However, because of security and strategic concerns, the NER of India appears hesitant to open up to trade with China. It is necessary to find out how one could minimize this security anxiety and establish a workable sub-regional mechanism to further the regional development in the coming future. This study, analysing the mode and structure of the economic development of Yunnan and BCIM-related areas in India, shows the possibilities of cooperation between Yunnan and the areas covered by BCIM in India. Proactive intervention by India and China would go a long way in the development of these two regions.

NOTES

1. Data from Office of Registrar General of India, Ministry of Home Affairs.
2. Calculated based on The Report of 'Basic Statistics of North Eastern Region 2015', North Eastern Council Secretariat.
3. See, *Global Times* (2015).
4. See, *The Hindu* (2016).
5. See, for more details, Ministry of Development of North Eastern Region, https://mdoner.gov.in/activities/rationale
6. See the North Eastern Development Finance Corporation Ltd (NEDFi) website: http://www.nedfi.com (accessed on 2018-11-13).
7. See, for more details, General Office of the People's Government of Yunnan Province.
8. See, *Xinhua News* (2012).
9. YuXinOu Railway is a freight rail route linking the south-western Chinese city of Chongqing with Duisburg, Germany. It passes through the Alataw Pass into Kazakhstan and moves through Russia, Belarus and Poland before arriving in Duisburg. The railway is part of a growing rail network connecting China and Europe along the New Silk Road.

REFERENCES

Aiyar, Pallavi. 2016. 'Developing Northeast the Yunnan Way'. http://www. thehindu.com/todays-paper/tp-opinion/Developing-Northeast-the-Yunnan-ay/article13113563.ece (accessed 2 November 2018).

Che Tailai. 2010. 'Development and Innovation of China-Europe International Logistics on Land Transportation (中欧国际物流陆路运输的发展与创新)'. *China Transportation Review* 5:33–39.

CPC. 2019. 'Reports of the 19th National Congress of the Communist Party of China' (习近平在中国共产党第十九次全国代表大会上的报告). Accessed

November 13, 2018. http://cpc.people.com.cn/n1/2017/1028/c64094-29613660-14.html.

DONER, RIS, and NIC. 2012. 'Ministry of Development of North Eastern Region, North Eastern Council: Expansion of North East India's Trade and Investment with Bangladesh and Myanmar: An Assessment of the Opportunities and Constraints'. https://www.ris.org.in/expansion-north-east-india%E2%80%99s-trade-and-investment-bangladesh-and-myanmar-assessment-opportunities (accessed 8 February 2021).

Downie, Edmund. 2014. 'Narendra Modi's Northeast India Outreach'. https://thediplomat.com/2014/12/narendra-modis-northeast-india-outreach/ (accessed 14 November 2019).

General Office of the People's Government of Yunnan Province. 'Interpretation on *Implementation Plan of Building Yunnan as an Economic and Trade Center Towards South Asia and Southeast Asia*' (云南省人民政府办公厅：解读《云南省建设面向南亚东南亚经济贸易中心实施方案》), http://www.yn.gov.cn/jd_1/jdwz/201703/t20170327_28887.html (accessed 8 February 2021).

Global Times. 2015. 'India Plans to Help Myanmar Build Border Roads (印度拟帮助缅甸修筑边境公路)'. http://china.Huanqiu.com/News/mofcom/2015-03/6031596.html (accessed 7 September 2019).

Gongbu, Duojia. 2018. 'A Study on the Tea-Horse Trade and National Cultural Communications (茶马古道商贸活动与民族文化交流研究)'. *Journal of Sichuan Minzu College* 1:1–7.

Hao, Chen. 2020. 'Yunnan Chapter Struggling to Compose the Chinese Dream'. *Qiu Shi* 2020 1 (陈豪.奋力谱写中国梦的云南篇章，《求是》，2020年1月).

Iyer, Roshan. 2017. 'A Promising Trilateral: India-Myanmar-Thailand'. https://thediplomat.com/2017/09/a-promising-trilateral-india-myanmar-thailand/ (accessed 13 October 2019).

Li, Min. 2011. 'A Comparative Study on Status Quo and Policy of Opening Up in Yunnan and Guangxi'. *Southeast and South Asian Studies* 3:9–16. (李敏：《云南与广西对外开放现状与政策对比分析》,《东南亚南亚研究》,2011年第3期。)

Liu, Sisi. 2014. 'BRI: A New Approach to Theoretical Research on Trans-regional Cooperation.' *South Asian Studies* 2:1–11. (柳思思："一带一路"：跨境次区域合作理论研究的新进路,《南亚研究》,2014年第2期。)

Lu, Xiaokun. 2019. *Analysis on Trade Cooperation Between Yunnan and India* (论云南与印度的贸易合作). Academic Exploration.

Lv, Zhaopyi. 2009. 'Strategic Change in Northeast India and Suggestions on Promoting Cooperation between Yunnan and Northeast India (印度东北地区的战略转变及推进中国云南与印度东北地区合作的建议)'. *Southeast and South Asian Studies* 4:17–29.

Maini, Tridivesh Singh. 2018. 'Eastern India's Embrace of China'. https://thediplomat.com/2018/01/eastern-indias-embrace-of-china/ (accessed 13 November 2018).

Nandy, S. N. 2014. 'Road Infrastructure in Economically Underdeveloped North-east India: A District Level Study'. *Journal of Infrastructure Development* 62:131–144.

People News. 2015. 'Yunnan: Gateway to South and Southeast Asian (云南：建面向南亚东南亚辐射中心)'. http://politics.people.com.cn/n/2015/0309/c70731-26660212.html (accessed 9 November 2018).

Rahmatullah, M. 2011. 'Strengthening Regional Transport Connectivity Among BCIM Countries'. *BCIM Newsletter*, Volume 1.

Raimedhi, Sriparna Pathak. 2015. 'Foreign trade and Northeast India'. http://www.orfonline.org/research/foreign-trade-and-northeast-india/ (accessed 14 November 2019).

Shengda, He, ed. 1993. *Contemporary Myanmar*, 322. Chengdu: Sichuan People's Publishing House.

The Hindu. 2016. 'India Begins Work on Bridge Linking Northeast to Chittagong'. https://www.thehindu.com/news/international/south-asia/India-begins-work-on-bridge-linking-northeast-to-Chittagong/article14416647.ece (accessed 13 October 2018).

The State Council. 2019. 'Expanding Domestic Demand and Increasing Investment to the Western Provinces (国务院:扩内需新增投资继续向西部倾斜)'. http://www.gov.cn/ (Accessed 16 November 2018).

Transport Department of Yunnan Government. (2019). 'Smooth Running at High Speed Way in Yunnan (高速畅行彩云南)'. http://www.ynjtt.com/Item.aspx?id=74148 (accessed 13 November 2018).

Varma, Subodh. 2009. '7 of 8 Northeast States Lag Behind Average India Income'. https://timesofindia.indiatimes.com/india/7-of-8-Northeast-states-lag-behind-average-India-income/articleshow/5187537.cms (accessed 13 November 2018).

Wang, Luolin, and Houkai Wei. 2003. 'The Progress and Evaluation of China's Development of Its West'. *Finance & Trade Economics* 10:23–32（王洛林，魏后凯：《我国西部大开发的进展及效果评价》,《财贸经济》,2003年10期）.

Xinhua News. (2012). 'Twenty Two New Key Projects Have Been Launched Under China's Western Development Project This Year (今年我国西部大开发新开工 22 项重点工程)'. http://news.xinhuanet.com/fortune/2012-12/19/c_114088238.htm (accessed 13 November 2018).

Chapter 7

The Northeast and Myanmar
India-Myanmar Engagement in the Modi Years

Munmun Majumdar

INTRODUCTION

India shares a 1,643-km-long international border with Myanmar. The land border traverses India's north-eastern states of Arunachal Pradesh, Manipur, Mizoram and Nagaland. Myanmar is geopolitically significant to India, as it links South Asia to Southeast Asia. It is the lone Association of Southeast Asian Nations (ASEAN) country that shares a border with India. It is also a member of the Mekong–Ganga Cooperation and the Bay of Bengal Initiative for Multi-Sectoral Technical and Economic Cooperation (BIMSTEC). Engaging Myanmar, therefore, becomes crucial for the success of India's Act East policy (AEP) that seeks to reach out to Southeast Asian and East Asian countries. The growing influence of China (Engh 2016) in Myanmar also compels India to take a greater interest in Myanmar. Further, Myanmar can provide the land-locked Northeast India a natural gateway to the sea and international market. These pragmatic considerations have compelled Government of India to strengthen its relations with Myanmar. The unveiling of the AEP at the 12th ASEAN–India Summit held in 2014 at Nay Pyi Taw in Myanmar (Ministry of External Affairs 2014), the extension of invitation to the president of Myanmar at the second swearing-in ceremony of Narendra Modi in 2019 (*Business Standard* 2019a) and the creation of a separate bureaucratic division for Myanmar in the Ministry of External Affairs (MEA) underscore the importance that India gives to its relations with Myanmar and the ASEAN. Against the background of the recent initiatives taken by Government of India, this chapter throws light on various issues of mutual concern that need to be addressed to strengthen the relations between India and Myanmar.

POLITICAL ECONOMY OF THE FREE MOVEMENT REGIME

The ethno-social fabric of the indigenous communities inhabiting the borderlands of India–Myanmar exercises considerable influence on the geopolitics of the region. Indo-Myanmar borderlands that house several indigenous ethnic communities influence the geopolitics in the region (Agnew 2019). Many of the indigenous tribes in India's North Eastern Region are ethnically linked to tribes on Myanmar's side of the border (Lall 2006). The borderland communities on both sides of the border view any effort to construct border fencing as inconsistent with the traditional lifeworld of the communities inhabiting the region. The Free Movement Regime (FMR) that both countries agreed to facilitates the ethnic groups residing along the border to travel 16 km across the boundary without visa restrictions. The FMR allows access to the border communities to interact with one another, maintain their age-old ties and engage in border trade. The barter system of trading, which was in practice for decades, has been replaced in recent years by cash transactions. Similarly, the development of road transport has eased the burden of head loaders carrying their products to border *haats* for sale. Establishment of banks on both sides of the border has eased monetary transactions.

The FMR is a strength and also a cause of concern for the security establishment, more so because the border passes through a region where many insurgents operate (Saikia 2007). The north-eastern states of India are plagued by ethnic conflicts, insurgent movements and weak infrastructures. Various insurgent groups are active at the border regions of India and Myanmar. Several armed groups operating in the region are involved in illicit narcotic and arms trade. The shared ethnic and cultural linkages enable these elements to obtain sustenance from both sides of the border. The insurgents have also been taking advantage of the FMR and have been crossing over to Myanmar. There has been a vast network of shadowy exchanges across the India–Myanmar border that has been in place since long before roads were built. These exchanges were outside state control and have occupied the realm of the shadow economy or informal economy. The volume of grey trade between India and Myanmar is much more than that of formal trade. India cannot ignore the ground realities while pursuing the AEP.

INDIA'S BUSINESS INTERESTS IN MYANMAR

The borderlands of Myanmar have become a hotbed of global and regional influences. China and India have critical national interests and long-term ambitions in Myanmar (Egreteau 2008, 2010), as Myanmar is a 'sleeping petroleum giant' (Ryon 2014). Rakhine is a borderland province having large oil and natural-gas reserves. It is necessary for India to present a commercially viable project for energy cooperation with Myanmar which would complement the bilateral efforts in building up infrastructure and improving connectivity. Constructive engagement with Myanmar would not only facilitate promotion of overland trade and transit routes and tourism but also allow room for counterterrorism, border management, etc.

Efforts have been stepped up by the Narendra Modi administration to take advantage of the long land border in order to expand cross-border trade, which stood at US$194.6 million in late 2019. On the economic front, India sees Myanmar as vital to fulfilling its ambition to become a US$5 trillion economy by 2024 (*Reuters* 2019). Much of the total trade at present happens through the sea (Sinate, Fanai and Bangera 2019). Myanmar's investments are small, whereas Indian investments in Myanmar through some 33 companies totalled about US$763.6 million in 2019 (Bilateral Brief). Most of India's investments have been in the oil and gas sector. Several Indian companies have also set up operations in Myanmar, including oil and gas players like the ONGC Videsh and GAIL. Banks, such as the State Bank of India, United Bank of India, now merged with Punjab National Bank, and Exim Bank of

India, have opened representative offices. India has been developing the bordering areas under India-Myanmar Border Area Development. Around the end of August 2020, India handed over US$5 million for the third year for the development of roads and bridges and schools undertaken in nine townships with 82 beneficial villages (*South Asia Weekly Report* 2020).

India's total trade with Myanmar, which was US$2,004.78 million in 2014–2015, has gone up to US$844,156.51 million in 2018–2019. An Integrated Check Post (ICP) at Moreh became functional from January 2019. The offer of tariff preferences to Myanmar under the Duty Free Tariff Preference (DFTP) scheme and conclusion of the ASEAN-India Trade in Goods Agreement are likely to have a positive impact on India–Myanmar trade. However, it should also be noted that despite expansion of trade volume over the years, the volume of trade across the India–Myanmar border remains relatively low in comparison with that across the Sino-Myanmar and Thai-Myanmar borders (Sinate, Fanai and Bangera 2019). Apart from ethnic insurgencies, the inadequate infrastructure facilities and logistics, such as roads, customs, ports, border *haats* and electricity, stand as obstacles to border trade. To increase the flow of goods between India and Myanmar, it is necessary to address infrastructure bottlenecks and constraints at the ICP and Land Customs Stations (LCSs). Given the importance of border trade, along with the connectivity plans in recent years, policymakers should give more attention to strengthen India–Myanmar economic relations.

CONNECTIVITY

India regards Myanmar as the first step to establish links with the rest of Southeast Asia. Therefore, it becomes imperative that India invest in connectivity and infrastructure. At present, there are four key connectivity projects underway. The first is the Kaladan Multi-Modal Transit Transport Project (KMMTT project), which is a corridor based on sea, river and road that aims at connecting the eastern Indian seaport of Kolkata with the Sittwe deep-water port in Myanmar's Rakhine State and seeks to improve India's access to its NER through Myanmar. The KMMTT project initiated in 2008 provides an alternative route through Myanmar for the transportation of goods to the NER of India. Considering its significance to the KMMTT project, India has been assisting Myanmar in developing Sittwe port, which is geostrategically located on the Bay of Bengal (Pattanaik 2019). The second project is the India–Myanmar–Thailand Trilateral Highway, a flagship land connectivity project between India and the ASEAN, announced in 2002. The Trilateral Highway connects India, Myanmar and Thailand through Moreh, Bagan and Mae Sot in the respective countries and is part of the Asian highway project. The Tamu–Kalay–Kalewa road, also known as the India–Myanmar Friendship Road, completed in 2001, is an arm of the Trilateral Highway project. Third is the Mekong–India Economic Corridor, which aims to jump-start India–Southeast Asia trade and investment linkages through connecting Chennai to Ho Chi Minh City through Dawei Port in Myanmar. BIMSTEC between the BBIN (Bangladesh, Bhutan, India, Nepal) countries, Sri Lanka, Thailand and Myanmar is the fourth project. The proposed BIMSTEC Motor Vehicle Agreement and BIMSTEC Master Plan on Transport Connectivity would give a further boost to the project.

As the ongoing Indian mega-projects will take a long time to be completed, India is now concentrating on smaller ones that are much more easily deliverable. India has been taking up development work under the Rakhine State Development Programme (RSDP) with a focus on agriculture and skill development. It announced a grant of USD 2 million for the construction of the border *haat* bridge at Byanyu/Sarsichauk in Chin State which would provide increased economic connectivity between Mizoram and Myanmar.

BANGLADESH–CHINA–INDIA–MYANMAR CORRIDOR AND CHINA'S INTERESTS

India and Myanmar are also signatories to the proposed Bangladesh–China–India–Myanmar (BCIM) corridor, which seeks to connect India and China through Myanmar and Bangladesh. The proposed corridor would connect China's Yunnan province to eastern India, traversing through Myanmar and parts of Northeast India and Bangladesh. The BCIM corridor proposals included greater access to goods, services and energy, elimination of tariff barriers, better trade facilitation, infrastructure development and joint explorations. This initiative of the Congress-led United Progressive Alliance (UPA) government could not take off as expected after Modi assumed power. Whereas China intended to make the BCIM corridor part of its Belt and Road Initiative (BRI), India wanted the BCIM corridor to be an independent project not linked to the BRI. Because of differences and mutual apprehensions between India and China, the BCIM corridor has not been able to make much headway.

Besides the BCIM corridor, India is facing problems in the implementation of other infrastructure projects too, compelling it to halt or slow down the pace of the connectivity projects. The factors such as difficult terrains, inclement weather conditions, logistical and systemic barriers, procedural issues, inadequacy of commercial and financial resources, insurgency, with its attendant risks, trans-border crimes, etc. affect project implementation. India's efforts to meet project deadlines are also affected by the continuing COVID-19 pandemic. Similarly, although the governments of both India and Myanmar welcomed the Memorandum of Understanding (MoU) between their respective private operators to launch a Coordinated Bus Service between Imphal and Mandalay by 7 April 2020, the proposal could not be implemented due to the pandemic.

Although the BRI-driven plan to link China and Myanmar does not directly have a bearing on India's connectivity projects with Myanmar (Rana 2020), the growing interdependency between China and Myanmar is a cause of worry for India. China is investing heavily in exploration of Myanmar's gas and oil resources in order to diversify its energy supply routes. The frequent visits of top leaders of China to Myanmar and conclusion of several agreements between the two countries indicate China's urgency to fast-forward its BRI projects through Myanmar. If China succeeds to push the BRI through Myanmar, it could ensure its energy supplies, maintain its economic growth and secure its military interests. China is developing a deep-sea port in Kyaukpyu in Rakhine State which would give it direct access to the Indian Ocean. If its plan materializes, China would get access to an alternate sea route, circumventing the Malacca Strait

(Renato 2015). China, through its investments in Myanmar, thus aims at having access to ports and waterways connecting it to the Indian Ocean. China's growing influence in the borderlands of Myanmar has serious geopolitical and strategic implications for India. Hence, India is compelled to strengthen its cooperation with Myanmar.

SECURITY COOPERATION AT THE BORDER

Border regions reflect continuity and change in the physical, political and social nature of state borders. Intercultural and mental borders can also have profound impacts on everyday life (Scott 2019). Economic development and national security are two sides of the same coin. For improving economic relations, India and Myanmar must boost their border security cooperation. It is only with a secure border that greater economic activity can take place. For example, there are risk factors that might adversely affect the KMMTT project, because it runs through the Chin and Rakhine provinces, where various armed movements from Myanmar, Bangladesh and India operate (Holslag 2009). The resistance posed at present by the Arakan Army (AA), for example, has put the future of the KMMTT project at risk. There have been several incidents of kidnapping of officers and workers engaged in the KMMTT project (PIB 2019), threatening the very survival of the project. The Myanmar Army—the Tatmadaw—is engaged in a battle of sorts with the AA rebels and the Arakan Rohingya Salvation Army. Myanmar's government declared AA as a terrorist group (Thura and Aung 2020). In 2019, India's and Myanmar's security forces undertook joint operations to push back the AA cadres targeting the Kaladan project (Bhalla 2020). The security situation in India's north-eastern borderlands is no better. Several native ethnic militant groups take refuge in Myanmar, taking advantage of the open borders (Bhaumik 2009; Nepram 2002; Upadhyaya 2019). Cross-border movement of insurgents is only one of several security challenges facing the policing of the Indo-Myanmar border. In recent years, realizing the security threats that the insurgent groups pose, India's and Myanmar's armies came together to carry out joint military operations, code-named Operation Sunshine-1 and Operation Sunshine-2, against the militants along the borders of Myanmar's Rakhine State that borders the north-eastern states of Arunachal Pradesh, Nagaland, Manipur and Mizoram.

The other areas of great concern to both India and Myanmar arise out of the rampant gunrunning and drug trafficking, which pose security challenges across the India–Myanmar border (Das 2019). Over the years, the India–Myanmar borderland has become the principal conduit for the trafficking of arms and heroin from Myanmar (UNODC 2010a, 137). Aligned with this problem is the spread of drug addiction and acquired immunodeficiency syndrome (AIDS) around the area, which is a safe haven for unbridled trafficking of women and children, smuggling of arms and gold, infiltration and cross-border movements of insurgents (UNODC 2010b, 41). Strict patrolling is almost impossible, owing to the hostile and harsh terrains creating instability, and consequently concerns are raised about the efficacy of the existing border security system. All these have made this once-neglected physical space a vulnerable zone from both traditional and non-traditional security dimensions.

To address this situation, India and Myanmar signed an MoU on border cooperation that provides a framework for security cooperation and exchange of information between India's and Myanmar's security agencies. The framework talks of coordinated patrols on their respective sides of the international border and the maritime boundary by the armed forces of the two countries. Both sides agreed to exchange information in the fight against insurgency, arms smuggling and drug, human and wildlife trafficking and agreed to take steps to prevent illegal cross-border activities. The level and frequency of meetings between the armed forces, drug control agencies and wildlife crime control agencies have been spelt out in the MoU.[1]

India and Myanmar have had director general–level talks to take coordinated and concerted actions against drug trafficking across their borders. The two sides have shared mutual concerns regarding the menace of drugs and resolved to exchange crucial information relating to the trafficking of drugs and precursors. They have been discussing the issues such as the trend of illicit poppy cultivation and heroin production in Myanmar, including precursor trafficking, cross-border trafficking of heroin from Myanmar to India, cross-border trafficking of ephedrine/pseudo-ephedrine, trafficking of methamphetamine from Myanmar to India, information on drug trafficking routes, controlled delivery operations and sharing of best practices and training needs (*Business Standard* 2019b). Both sides also agreed to step up intelligence cooperation to fight insurgency along the border with Bangladesh, north-eastern states and China—where insurgents are active.

DEFENCE AND MARITIME COOPERATION

Following the then Army Chief General V. P. Malik's Myanmar visit in January 2000, military ties between India and Myanmar got further strengthened. A military agreement was sealed in July 2017 which provided for counter-insurgency operations in India's Northeast and training of Myanmar's armed personnel in defence training establishments. The visit of Senior General Min Aung Hlaing, Commander-in-Chief of Defence Services of Myanmar from 25 July to 2 August 2019, once again echoed the enhancement of bilateral defence relations between India and Myanmar (Lwin 2019). The signing of the Defence Cooperation Agreement marked the deepening of their defence relations. The meeting between Minister of State for Defence Sripad Yesso Naik and Senior General Min Aung Hlaing is important in the context of the joint operations against insurgent groups from India's Northeast active in the borderlands of Myanmar.

Narendra Modi's visit to Myanmar in 2018 contributed to expansion of the security and defence partnership. The armed forces of the two countries have been engaged in several training exercises and drills. India provides military training and conducts joint military exercises with the Myanmar Army. Through the India–Myanmar bilateral military exercises (IMBAX 2017 and IMBEX 2018–2019), India trained the Myanmar Army to take part in United Nations peacekeeping missions. Myanmar has been a participant in the Milan naval exercises hosted by the Indian Navy since 2003. The navies of India and Myanmar carried out the India–Myanmar Naval Exercise in 2018 and 2019. Defence cooperation with India enables Myanmar to diversify its defence imports. Myanmar purchased India's first locally produced anti-submarine torpedo called

Torpedo Advanced Light (TAL) Shyena in 2017, and in 2019, Myanmar acquired a diesel–electric Kilo-class submarine called INS Sindhuvir, which India modernized after buying it from Russia in the 1980s. For Naypyidaw, these military purchases are meant to secure and protect Myanmar's maritime interests, especially after its neighbours Bangladesh and Thailand acquired submarines from China.

SOFT-POWER DIPLOMACY

India has been exploring the efficacy of soft power (Nye 2004) in achieving foreign policy goals in Myanmar. Prime Minister Narendra Modi on his first trip to Myanmar invoked India's historic and cultural ties with Myanmar to build a strong and sustainable relationship. In keeping with the MoU for the conservation of earthquake-damaged pagodas at Bagan (Roy Chaudhury 2018), India has been assisting the preservation of cultural heritage, including the repair and conservation of the Bagan pagoda. The two sides welcomed the commencement of the first phase of work by the Archaeological Survey of India (ASI) to restore and conserve 12 pagodas under a project to restore and conserve 92 earthquake-damaged pagodas in Bagan. India also undertook conservation and renovation of stone inscription temples and Zayat of King Bagyidaw and King Mindon at Bodh Gaya, with full funding from the MEA. India lent support to Myanmar in improving the socio-economic condition of the people in Rakhine State. As part of the RSDP, India is actively associated with Phase-III of the programme and helped in setting up a skills training centre. After Prime Minister Narendra Modi's visit to Myanmar in 2017, there have been discussions exploring various infrastructural and socio-economic projects in Myanmar.

India has been assisting Myanmar in improving education, training and skill development. Indian Institute of Information Technology (IIIT) in Bangalore and Myanmar Institute of Technology have entered into a close collaboration. Private players, such as Tata Consultancy Services (TCS), have been undertaking projects in Myanmar as well. India has provided loans to develop and upgrade infrastructure projects, such as the Yangon–Mandalay rail link, and in the recent past, as part of the AEP, the Narendra Modi government has begun negotiations for upgrading road links between the two countries. Further, as in the past, India offered medical radiation equipment Bhabhatron 2 for treatment of cancer patients, and both sides agreed to further strengthen cooperation in the healthcare sector. India also agreed to deepen cooperation to overcome the challenges posed by the COVID-19 pandemic. Symbolic of India's commitment to assist Myanmar, India also presented 3,000 vials of Remdesivir to Myanmar's State Counsellor.

In order to promote tourism and attract foreign tourists and pilgrims from Southeast Asian countries, Prime Minister Narendra Modi proposed the ambitious Buddhist Circuit initiative that seeks to connect ancient Buddhist heritage sites across different states in India. The initiative could bolster India's tourist industry through the 'Incredible India' campaign and consolidate India's diplomatic reservoir of goodwill with the ASEAN countries. Similarly, India can explore shared ties and cultural similarities among the people living in Indo-Myanmar borderlands to strengthen the relations between the two countries. Being aware of the fact that political stability in Myanmar is essential to maintaining India's security along the border, Narendra Modi's government has been supporting the effort to achieve peace and reconciliation in Myanmar and has participated as a signatory witness to Myanmar's Nationwide Ceasefire Agreement ceremonies.

During their visit to Myanmar in October 2020, Chief of the Indian Army Staff Gen Manoj Naravane and Foreign Secretary Harsh Vardhan Shringla held talks with high-level officials in Myanmar. Both sides discussed the progress of infrastructure and connectivity projects and agreed to further strengthen their partnership in capacity building, power and energy and deepen economic and trade ties. Further, they also agreed to facilitate people-to-people and cultural exchanges and broad-based defence exchanges across all the three military services. Reiterating their mutual commitment not to allow their respective territories to be used for activities inimical to each other, both sides also discussed the maintenance of security and stability in their border areas. They also agreed to deepen cooperation to overcome the challenges posed by the COVID-19 pandemic. They exchanged views on an early initiation of work on fresh initiatives, such as the upgradation of Yamethin Women's Police Academy and Basic Technical

Training School. They agreed to work towards operationalization of Sittwe Port in Rakhine State in the first quarter of 2021.[2] The two sides discussed plans to install a bust of Lokmanya Tilak in Mandalay to commemorate his 100th death anniversary. Other areas of cooperation in culture discussed included translation of Indian epics (Ministry of External Affairs 2020). The virtual inauguration of the Centre of Excellence in Software Development and Training in Myitkyina and the Embassy Liaison Office in Nay Pyi Taw, along with that of U Soe Han, Permanent Secretary, Ministry of Foreign Affairs of Myanmar, are cumulative pointers to the successful outcome of India's Myanmar engagements. Another major takeaway from the visit was India's proposal to build a US$6 billion petroleum refinery project near Yangon. This would establish India's credentials as a sustainable development partner, at the same time widening the scope for Myanmar to diversify.

CONCLUSION

In recent decades, India has been making efforts to strengthen its relations with Myanmar. Although the process was initiated during the UPA regime, the momentum picked up after Narendra Modi became the prime minister. During the NDA rule, India has made commendable strides in improving bilateral ties and explored new avenues of cooperation with Myanmar. India realizes that Myanmar has been crucial in the pursuit of its own national interests. Ensuring the security of the Indo–Myanmar border is essential for maintaining law and order, and for initiating economic development in the NER. As Myanmar is a gateway to Southeast Asian countries, the success of India's Look East policy (LEP)/AEP depends to a considerable extent on how effectively India engages its neighbour Myanmar. China's growing influence in Myanmar is also compelling India to take proactive initiative in the region. Over the years, there has been an expansion in the volume of trade between India and Mayanmar. India has invested in energy and connectivity projects in Myanmar. Its soft power diplomacy is likely to pay dividends in the long run. Peace, stability and economic development in Myanmar are vital for India as well. Hence, India has to tread a cautious and pragmatic path that builds mutual trust, goodwill and cooperation between the two countries.

NOTES

1. India and Myanmar signed 'Memorandum of Understanding on Border Cooperation', Ministry of External Affairs, Government of India. https://www.mea.gov.in/press-releases.htm?dtl/23315/India+and+Myanmar+sign+Memorandum+of+Understanding+on+Border+Cooperation (accessed 10 May 2014).
2. See *Hindustan Times* (2020).

REFERENCES

Agnew, J. 2019. 'Dwelling Space Versus Geopolitical Space: Reexamining Border Studies in Light of the Crisis of Borders'. In *Debating and Defining Borders: Philosophical and Theoretical Perspectives,* edited by A. Cooper and S. Tinning, 57–69. London: Taylor & Francis.

Bhalla, Abhishek. 2020. 'Supply of Chinese Arms to Myanmar's Arakan Army Threatens Safety of Kaladan Project Vital for NE India'. *India Today*, 28 October. https://www.indiatoday.in/india/story/myanmar-arakan-army-kaladan-project-northeast-india-mizoram-1735992-2020-10-28 (accessed 8 February 2021).

Bhaumik, Subir. 2009. *Troubled Periphery: The Crisis of India's North East.* New Delhi: SAGE Publications.

Bilateral Brief. Embassy of India Yangon, India-Myanmar. https://embassyofindiayangon.gov.in/ (accessed 22 October 2020).

Business Standard. 2019a. 'Myanmar Prez Reaches Delhi to Attend Modi's Swearing-in Ceremony'. 30 May. https://www.business-standard.com/article/news-ani/myanmar-prez-reaches-delhi-to-attend-modi-s-swearing-in-119053000478_1.html (accessed 8 February 2021).

———. 2019b. 'India, Myanmar Begin Talks to Take Coordinated Efforts Against Drug Trafficking'. 9 July. https://www.business-standard.com/article/news-ani/india-myanmar-begin-talks-to-take-coordinated-efforts-against-drug-trafficking-119070901274_1.html#:~:text=India%2C%20Myanmar%20begin%20talks%20to%20take%20coordinated%20efforts%20against%20drug%20trafficking,-ANI%20%7C%20Asia%20%7C%20Last&text=%22The%20two%2D%20day%20bilateral%20meeting,countries%2C%22%20the%20statement%20added (accessed 8 February 2021).

Das, P. 2019. 'Security Challenges and the Management of the India–Myanmar Border'. *Strategic Analysis* 42(6):578–594.

Egreteau, R. 2008. 'India and China Vying for Influence in Burma—A New Assessment'. *India Review* 7(1):38–72.

———. 2010. 'India's Unquenched Ambitions in Burma. In Burma or Myanmar?' In *The Struggle for National Identity*, edited by Lowell Dittmer, 295–326. Singapore: World Scientific.

Engh, S. 2016. 'India's Myanmar Policy and the Sino-Indian Great Game'. *Asian Affairs* 47(1):32–58.

Hindustan Times. 2020. 'India, Myanmar Discuss Operationalisation of Sittwe Port, Indian Side Announces New Initiatives'. New Delhi, 5 October. https://www.hindustantimes.com/india-news/india-myanmar-discuss-operationalisation-of-sittwe-port-indian-side-announces-new-initiatives/story-gqnSFtu6qjAB-GzxK5i33AP.html (accessed 8 February 2021).

Holslag, J. 2009. 'The Next Security Frontier: Regional Instability and the Prospects for Sino-Indian Cooperation'. *Strategic Analysis* 33(5):652–663.

Lall, M. 2006. 'Indo-Myanmar Relations in the Era of Pipeline Diplomacy.' *Contemporary Southeast Asia* 28(3):424–446.

Lwin, N. 2019. 'India, Myanmar Strengthen Military Ties'. *The Irrawaddy*, 30 July. Myanmar. https://www.irrawaddy.com/news/burma/india-myanmar-strengthen-military-ties.html (accessed 4 June 2020).

Ministry of External Affairs. 2020. 'Visit of Chief of Army Staff and Foreign Secretary to Myanmar'. Government of India, 4–5 October. https://www.mea.gov.in/press-releases.htm?dtl/33092/Visit+of+Chief+of+Army+Staff+and+Forei gn+Secretary+to+Myanmar+October+45+2020#:~:text=Partnerships%20 Development%20Partnerships-,Visit%20of%20Chief%20of%20Army%20 Staff%20and%20Foreign%20Secretary%20to,October%204%2D5%2C%20 2020)&text=Chief%20of%20Army%20Staff%20(COAS,October%20 4%2D5%2C%202020 (accessed 8 February 2021).

Nepram, Binalakshmi. 2002. *South Asia's Fractured Frontier: Armed Conflict, Narcotics & Small Arms Proliferation in India's Northeast*. New Delhi: Mittal Publications.

Nye, S. Joseph. 2004. *Soft Power: The Means to Success in World Politics*. New York: Public Affairs.

Ministry of External Affairs. 2014. Opening Statement by Prime Minister at the 12th India-ASEAN Summit, Nay Pyi Taw, Myanmar, 12 November. http://mea.gov.in/aseanindia/SpeechStatementASEM.htm?dtl/22566/ Opening+Statement+by+P rime+Minister+at+the+12th+IndiaASEAN+S ummit+Nay+Pyi+Taw+Myanmar (accessed 15 December 2019).

ORF. 2020. *South Asia Weekly Report*, Volume XIII–36, 8 September. https:// www.orfonline.org/research/south-asia-weekly-report-volume-xiii-36/ (accessed 22 October 2020).

Pattanaik, S. S. 2019. 'Geo-strategic Significance of Bay of Bengal and Andaman Sea: Leveraging Maritime, Energy and Transport Connectivity for Regional Cooperation.' *South Asian Survey* 25(1–2):84–101.

PIB. 2019. 'Government of India Secures Release of Five Indian Nationals Abducted in Myanmar'. 5 November. https://pib.gov.in/PressReleaseIframePage. aspx?PRID=1590382 (accessed 15 April 2020).

Rana, K. S. 2020. 'Conclusion: An Indian Perspective on the Belt and Road Initiative'. In *China-India Relations: Geo-political Competition, Economic*

Cooperation, Cultural Exchange and Business Ties, edited by Kim Young-Chan, 203–228. Cham: Springer.

Renato, Etac. 2015. 'The Malacca Dilemma: A Hindrance to Chinese Ambitions in the 21st Century'. *The Review*, September 23.

Reuters. 2019, November 14. 'Modi Says Wants to Make India a $5 trillion Economy by 2024'. https://www.reuters.com/article/us-brics-summit-india/modi-says-wants-to-make-india-a-5-trillioneconomy-by-2024-idUSK-BN1XN2RT (accessed 23 October 2020).

Roy Chaudhury, Dipanjan. 2018. 'PM Narendra Modi's visit to expand strategic and economic footprint in Myanmar'. *Economic Times*, 13 July. https://economictimes.indiatimes.com/news/defence/pm-narendra-modis-visit-to-expand-strategic-and-economic-footprint-in-myanmar/articleshow/60298588.cms?utm_source=contentofinterest&utm_medium=text&utm_campaign=cppst (accessed 8 February 2021).

Ryon, R. E. 2014. 'Foreigners in Burma: A Framework for Responsible Investment'. *Pacific Rim Law & Policy Journal* 23(3):832–867.

Saikia, J., ed. 2007. *Frontier in Flames: North East India in Turmoil*. New Delhi: Penguin.

Scott, W. J. W. 2019. 'Border Regions'. In *The Wiley Blackwell Encyclopedia of Urban and Regional Studies*, edited by A. M. Orum. New Jersey: John Wiley & Sons.

Shawn, W. Crispin. 2004. 'Pipe of Prosperity.' *Far Eastern Economic Review*, 19 February.

Sinate, D., V. Fanai, and S. Bangera. 2019. 'India-Myanmar Trade and Investment: Prospects and Way Forward' (Working Paper No. 90). Export-Import Bank of India. https://www.eximbankindia.in/Assets/Dynamic/PDF/Publication-Resources/ResearchPapers/110file.pdf (accessed 8 February 2021).

Thura, Myat, and Sit Thet Aung. 2020. 'Myanmar Govt Declare AA as a Terrorist Group'. *Myanmar Times*, 23 March. https://www.mmtimes.com/news/myanmar-govt-declares-arakan-army-terrorist-group.html (accessed 15 April 2020).

Upadhyaya, Archana. 2019. *India's Fragile Borderlands: The Dynamics of Terrorism in North East India*. New Delhi: Routledge.

UNODC. 2010a. *World Drug Report 2010*, 137. Vienna: United Nations Publication. https://www.unodc.org/documents/wdr/wdr_2010/world_drug_report_2010_lo-res.pdf (accessed 6 November 2020).

———. 2010b. *World Drug Report 2010*, 41. https://www.unodc.org/documents/wdr/wdr_2010/world_drug_report_2010_lo-res.pdf (accessed 6 November 2020).

Chapter 8

Trade in Pre-colonial Arunachal Pradesh and Tibet

Amrendra Kumar Thakur

Several researchers have studied the interconnections among various aspects of state, society and trade in pre-colonial Northeast India. They have also paid critical attention to economic and cultural contacts that existed among the tribes in the region (Thakur 2014, 2016). The scholars, using archaeological and literary sources, have illustrated the significant role played by Buddhist centres in the establishment and functioning of trade networks. The studies have examined the problems of transition between the pre-colonial and colonial periods and the resultant break in trade practices during the colonial period. Extending the research on similar lines, the present chapter attempts to take a fresh look at the trade practices of the tribes of Arunachal Pradesh with the people of the neighbouring areas in Tibet (now China). The chapter also provides a critique of the colonial conceptions of so-called primitivism and isolation of the tribes of Northeast India.

The colonial discourse on the tribes of Arunachal Pradesh is based on official colonial records and colonial ethnographic works. The recurrent and apparently the most-relied-upon colonial sources for understanding this tribal territory have been the detailed accounts of visits and tours of colonial officials to the tribes, most of which were written in an ethnographic style and claim to offer accurate knowledge of the tribal territories. These writings painted the native tribes as 'barbarous', 'savages' and 'wild men' who lived as cultural isolates in terra incognita, etc. To the European eyes, the tribal people were nothing but 'queer and exotic'. Since the West's knowledge of the East cannot be fully appreciated without identifying the political context within which the East became an object of that knowledge, the colonial construction of tribes as isolated and their social processes as static has to be viewed within the broader contexts of the state's search for order and its attempts to locate its unfamiliar within a known frame of reference.

The economic dependence on sociocultural relationship with and accommodation of differences of one village/community from the surrounding others have not been adequately highlighted, whereas their conflicts and disputes have been given greater importance in the studies of colonial writers. The inherent weakness of colonial writings is clearly evident as late as 1945. One officer-cum-author in one report mentions:

> Till I visited, … no officer had ever been known to enter a tribesman's house. This astonishing state of affairs illustrates the aloofness of successive P. O.'s [sic] [Political Officers] from their people and partly accounts for the almost complete ignorance. I found … this ignorance will in future be somewhat remedied, but the present P. O. [Political Officer], like the predecessors, has neither the aptitude nor taste for detailed enquiries into tribal customs and economics.[1]

The opinion of L. A. Waddell is the representative opinion of the colonial writers. He writes, 'Driven into these wild glens by the advance of civilization up the plains and lower valleys, these people have become hemmed in among the mountains, where pressing on each other in their struggle for existence, they have developed into innumerable isolated tribes'.[2] Contrary to the view of the colonial scholars, it has been found that the earliest reference to commercial relations between India and China through the Assam–Burma routes is found in the accounts of Changkien (200 BC) (Choudhury 1988, 272). *Arthashastra* of Kautilya shows commercial contacts between the Northeast and other parts of India, along with China (Mukherjee 1992, 12). Assam had been maintaining its trade contact with Tibet and Bhutan too. There were direct trade routes, through the mountain passes of the north, with Tibet and Nepal, extending up to the borders of Kashmir (Sarma 1981, 138). *Tabaqat-i-Nasiri* mentions that there were 35 passes between Assam and Tibet and that through them horses were brought to the kingdom of Kamarupa (Acharya 1972, 41; Choudhury 1987, 356–357). The people in the Assamese plains and the hills lying along the northern border maintained a steady trade contact with Tibet. There were many mountain passes between Bhutan and Tibet. The Tibetan traders entered the lower lands through various routes that opened on the *duwars* or gateways (connecting the foothills and the hills) like Bijni and Chapaguri in the present Kamrup district and Borigumma in the Darrang (Neog 2008, 89). The Bhutias used to travel through the valley of Manas River via Tasgong and Dewangiri to Hajo, where a fair was usually held.[3] The name Suvarnakudya is mentioned in *Arthashastra* as a place where fine silk cloths were produced. The place is identified by scholars as Hajo in the district of Kamarupa.[4] The Monpa and Sherdukpen communities, which were followers of Buddhism, also regard Shingri (Dhekiajuli of Assam) as their pilgrimage site along with Hajo.[5] Hajo, at a distance of 20 km from Guwahati on the north bank of the Brahmaputra, is one of the pilgrim centres of Assam. It is usually visited by Buddhists from Bhutan and Tibet. The sanctity accorded to Hajo becomes evident from the fact that the first monastery in Tibet was built with clods of earth collected from Hajo (Neog 2008, 36–37). Thus, pilgrimage centres, along with trade centres, were developing rapidly in the pre-colonial period in Northeast India.

The significant trade between Assam and Tibet through Arunachal Pradesh and the concurrent rise of Buddhism have received considerable scholarly attention. Historians and other social-science researchers have highlighted the crucial aspect of the participation of Buddhist monks in trade (Thakur 2012). The hierarchical monastic order of Lamaism, for instance, consolidated itself in northern Assam even as it originated in Tibet. Unsurprisingly, the same Lamaism can be noticed among the Sherdukpen, Monpa, Memba and Khamba of Arunachal Pradesh. Even the Bhutias who live along the border of western Arunachal Pradesh in Bhutan observe the same practice. In fact, the Monpas, Sherdukpens and Membas believe that Padmasambhava visited their territories in the 8th century CE and preached Buddhism. N. Sarkar refers to monastic records that confirm the construction of Buddhist monasteries in the 12th century CE or even earlier in these areas (Bhattacharya 1995, 133–136).

The Monpa and the Sherdukpen played important roles as intermediaries in trade between Assam and Tibet. They used to go to Tibetan trading marts during the summers and the Assamese plains during winter. The produces such as chillies, vegetables dyes, like madder, peaches, handmade paper and husked rice were traded for Tibetan salt, wool and *churpi* (frozen milk cake). The traded items of the Monpa, such as masks, animal hide and chillies, were exchanged with Tibetan rock salt, wool, woollen clothes and *dao* (iron axe or hatchet). Besides these, the local produce such as madder, red dye, tobacco, herbs and madder dye were taken to Tibet in exchange for symbols, Tibetan religious bells and white shells and beads. The Sherdukpen are also known for conducting trade among tribes of Arunachal Pradesh and with the people of Assam (Cosh 1837, 66). The annual migration of the Sherdukpen (to avoid the cold in their habitat in Arunachal Pradesh) to Doimara helped them trade in the plains of Assam. They used to take their cattle, poultry, chillies, dried radishes, *jabrang* (local pepper), *daos* and woven bags to Doimara for sale in the markets. They used to purchase Assamese *endi* clothes, salt, rice, beads, bangles and metal utensils. Within Arunachal Pradesh, the Sherdukpen used to exchange cows, Assamese *endi* clothes, *mithuns* and animal skins with the Monpa for butter, coats, shoes, carpets, blankets, masks and yak caps, and with the Khowas–cloths, salt and betel nut for other indigenous goods.

Besides the trade between Assam and Tibet through the frontier tribes, there was also some direct trade and communication between the two regions. In 1837, J. M. Cosh reported:

> During the flourishing period of the Assam dynasty, we [the author and his interpreters] are informed, that the kings of Assam were in the habit of sending presents to Grand Lama and that a caravan consisting of about 20 people annually resorted from Lassa to the Assam frontier; and transacted merchandise to a very considerable amount with the Assamese. The Thibetans [sic] took up their quarters at a place

called Chouna, two months' journey from Lassa and the Assamese, at Geganshur, a few miles distant from it. The trade at the former comprised silver and a large quality of rock salt. This they exchanged with the Assamese for rice, silk, lac and other produce of Bengal....[6]

Unlike Tibetan trade, the trade with Bhutan appears to have been limited to some pockets of western Arunachal Pradesh. Bhutanese trade mostly comprised articles of Tibetan origin. Since the people of western Arunachal Pradesh not only had direct access to Tibetan markets but also shared ethnic and religious affinities with the Tibetans, Bhutanese trade never featured in the state of Arunachal Pradesh. It has been expressed during my field studies that a few articles of Bhutanese origin—especially the *tonga*, a coarse woollen cloth—which were popular among the frontier tribes found their entry into Arunachal Pradesh through Tibet.

The people, especially of the central and western zones, living in the high ridges (the Memba, Khamba and Monpa of Mago–Thingbu areas) had a very meagre scope for cultivation. Therefore, trade was their principal occupation, and it helped maintain symbiosis with the economy of their neighbouring communities. Here, their geographical and cultural proximity with Tibet proved to be very important. There was trade between Tibet and the village of Shyo in Tawang district in pre-colonial and also post-colonial periods. It is important to mention that trade was very popular among not only the people of high mountainous regions but also those at the foothills. The discovery (fortunate but totally unexpected) of about 250 silver coins of different sizes, belonging most probably to the rulers of the Shah dynasty of Bengal (official report not yet published), in a ruined fort complex (Bhalukpong) at the foothills of western Arunachal Pradesh gives important insights into the study of the economic history of the region.[7] Most probably, the fort served as the central place of contact between traders of Bengal and Assam, on the one hand, and those of Arunachal Pradesh and Tibet, on the other.

The trade in slave also formed an important part in the trade relations of the people of Arunachal Pradesh with Tibet. During the (pre-) colonial period, Tibetan slaves were found in Arunachal Pradesh. Slave masters from Arunachal Pradesh also sold off their slaves in Tibet. The efforts made by the government to discourage the practice of slavery probably compelled the slave masters to dispose of the slaves to the neighbouring country (Tibet).[8]

Before colonization, the tribal chiefs in the region played important roles in building trade linkages between Arunachal Pradesh and Tibet. Apart from trading in goods directly, they acted as facilitators for development of a concomitant economy that relied on the collection of tributes and taxes. The pre-colonial trade between Arunachal Pradesh and Tibet was aided and supplemented by the economy of tax collection that the tribal chiefs sustained. The Bapus (chiefs, also referred to as rajas or kings) of Thembang and Namshu in Bomdila area, for instance, conducted a considerable amount of trade with the people of Mago area, from whom they collected tribute. As there was almost no cultivation in Mago, the people lived by trade in salt. Their sustenance, to some extent, was aided through exchange of surplus produce of livestock—herds of yak and sheep—for grains, chillies, madder dye and skins of wild game with the Bapus and the common people of Thembang and Namshu. They also traded, though rarely, with Lamai (Mijis) of the Dinam and Pachuk valleys. Every year, they received a considerable quantity of salt from Tibet in exchange for some articles they got from the Thembang, Namshu and Miji areas. Chillies and madder dye, which did not grow in Tibet, were in great demand there, and the amount of chillies and madder dye that found their way annually to Tibet via Mago was colossal.[9] As per the arrangement, the tribute-paying houses of Mago areas were grouped and allotted to each of the seven tribal chiefs (four from Thembang and three from Namshu). Each house had to pay annually either one yak or 13 *tankas* (silver coins) to the concerned chief and in return received 60 *bres* (*bre* is a unit of mass, approximately equivalent to 0.5 kg) of chillies or six *tankas* from the payee. In addition, the entire villages of Thembang and Namshu received tributes annually from the Mago and Lungthang areas. The tribute was paid every second year to Thembang and every third year to Namshu.

Like the relationship between the chiefs of Thembang and Namshu and the villages of Mago area, the villages of Phudung area (Phudung, Bamrok and Khelong) depended upon the Thongs (chiefs, also referred to as rajas or kings) of Rupa and Shergaon. The villagers of the Phudung area used to come to Doimara via Domkho and Senthui during winter months to exchange chillies and *jabrang* (a kind of spice for making curry) for paddy, salt and *endi* cloth with the Thongs and also to carry loads for the latter on payment in kind in the form of grain and salt. Here, it is important to state that despite their village being on the main Dirang Dzong–Amaratulla–Udalguri road and although subject to Talung Dzongpens, they hardly went to Amaratulla and Udalguri and depended upon the goods from the Thongs.

The 1946 'Report on tribute, other than monastic tribute, in the Se La Subagency' by I. Ali, Esq., Political Officer, also attests to the existence and efficacy of the trade relations among the tribes of Arunachal Pradesh which involved exchange of Tibetan goods for local ones. Ali, at the conclusion of the report, writes:

> In my view it would be most opportune to make an announcement prohibiting all payment and receipt of tribute, simultaneously with the introduction of house-tax in the Se La sub-agency, among the Monba and the Senjithongji…Of course it will be equally necessary to take every step to restore legitimate trade between the tribes, which has now come to stalemate, owing to stoppage of tribute and I am confident that with the extension of control, it would be possible.[10]

The Aka chief used to get tributes from the Thongs of Rupa and Shergaon areas. Trade also played a significant role in their relation. As per the tradition, the Aka chief Neymachongdi paid a visit to Thong village to renew his friendship with Gyaptang (the founder of the Thong chiefdom). The Aka chief grew jealous of him, having seen Gyaptang prosper out of the trade he had with the plains. Gyaptang was clever enough to win over the Aka chief by promising to present him a full-grown bullock, a large *endi* (silk) cloth and a sheep once in every 5 years. The Aka chief Neymachongdi was pleased, and in return, he assured that the Akas would cause no trouble while Gyaptang was away in the plains. Gyaptang also received some dried fish, dried venison, potatoes and taros from the chief.

It was reported that most of the chiefs who received land tributes also facilitated trade and participated directly in trading activities. They used to collect tributes on trade. Also, traders mostly traded themselves, received tributes on taxes or traded in their own favour. It was reported during field studies that the Tibetan government used to grant the monopoly over sale of salt to a contractor (the Dekangpa) in Tawang. He sold salt to every villager, most of whom paid in kind. He also enjoyed some privileges in getting free labour from the villages. Besides the salt monopoly, the contractor was also empowered to trade in rice. Any person buying rice from anyone had to exchange rice with the Dekangpa for salt. The violators, if caught, were punished.[11] It is also reported that people of the Dirang area were almost entirely dependent on the salt imported from Tsona by the official traders of Tawang monastery and that the representatives of the Tsona Dzongpons visited the villages in this area periodically to collect tribute and exchange salt for grains. Salt was rather expensive, as the villagers had to part with 10 *tres* (a Tibetan measuring unit, almost equal to 0.5 kg) of grain for 1 *tre* of salt. Some salt was also obtained from Mago. It was also informed that some Mijis also got their salt from Mago, exchanging it with their paddy, deer skin and musk. The Bhutanese from Tashi Dzong area used to visit Mago for trade. The Monpa living to the east of Dirang got a considerable amount of paddy, some millet, corn, dyes and deer skins from the Mijis, and in return the latter received Tibetan salt, blankets and some Tibetan necklaces. Very few of the Rahung and Kudam villagers came down to the plains for trade. Those who did took the Doimara route via Rupa.[12]

The influence of the pre-colonial political power structure in trade continued even during the 20th century. The *Tour Diary of Political Officer, Balipara Frontier Tract* clearly mentions, 'They (Dirangdzong) had also lent the use of an excellent commodious house to store the rations. The house is generally used by the Talungdzongpons and their traders on their way to and back from Amratulla'.[13] Besides facilitating trade, the Dzongpens (the officers in charge of the area) also engaged in trade—either directly or through their representatives. However, the trade was conducted on behalf of the monastery. The Dzongpens brought Tibetan salt, blankets and ornaments, which they exchanged for grains. Trade was always in their favour. All the grains collected, whether by way of monastic tribute or obtained in trade, had to be carried free by the villagers.

During the pre-colonial period, besides favourable trade, the states used to collect taxes from the traders as toll tax too. The matter has been reported in the tour diary mentioned above. There, it is stated:

> At Amratulla I saw the toll gate of the Dzongpens in which tolls (Chug) have to be paid by every trader going down to Udalguri in cash or kind and on return again everyone has to bring paddy from the plains (Assam). I saw some paddy thus collected being husked into rice and a good stock of it ready for dispatch to Tawang.[14]

The jurisdiction for this toll was the rite of passage given to the traders through the route from Tawang to Amaratulla, which they claim as theirs. There were also trade toll gates at Rupa and Shergaon, and they too collected tolls from traders using their route while coming down to Doimara. The Sherdukpen chiefs (Thongs) used to justify this toll saying that they had to encounter a good deal of trouble in clearing the paths from Rupa and Shergaon to Doimara every year and hence the people using it must pay something in return.

Paddy that they produced in excess of consumption was usually exchanged in trade. Even at a much later period, enormous quantities of rice were exported to Tibet in exchange for rock salt. In the inscriptions of Vanamaladeva, the water of the River Lauhitya is described as being polluted with the mud of gold that comes from the huge gold rock of Mount Kailash.[15] Gold washing was an important industry in Assam from which the government earned considerable revenue even in later days (Barua 1966, 121).

Thus, we can say that pre-colonial Arunachal Pradesh had highly developed internal and external trade. The volume of trade was very high. In 1809, the volume of trade through Doimara and Udalguri fairs amounted to ₹2 lakh (Hamilton 1940, 73–74). The protracted troubles of Assam caused because of the decline of the Ahom monarchy ultimately affected the trade traffic, but even in the year before the Burmese invasion, the Lhasa merchants were reported to have brought gold worth ₹70,000 (Mackenzie 1979, 15). At the Udalguri fair, the principal items of export from the Monpa areas were horses and ponies, gold, blankets, salt, musk, wax, spices, yak tails and rubber. The main items of import into Arunachal Pradesh were Assamese silk manufactures, brass and copper manufactures, iron, tobacco and betel nut. The common items of export from the Sherdukpen and Thembang Monpa areas at the Doimara fair were spices, wax, musk, chillies, salt, sheep, blankets, etc. The items of import into the various areas of Arunachal Pradesh were almost the same as those from Udalguri. An observation of the import and export items at these two fairs clearly shows that animals, animal products, forest-based resources and Tibetan goods made up the chief items of export from the western zone of Arunachal Pradesh, whereas the items imported into the zone constituted mainly of finished goods. Here, technological know-how, availability of natural resources in the forms of forests and minerals and fashioning of queer and exotic objects of Tibet and Assam seem to be the determining factors in trade.

The above discussion also throws sufficient light on an extensively existing and well-practised material culture in Arunachal Pradesh during the pre-colonial period. Based on the same, a detailed study would be a significant contribution. The study could deal with the drawing up of maps of zones and sub-zones of selected materials and cultural traits that appear to have persisted over long historical periods. The study of language, religion and religious sects, architecture, physical appearance (race), climate, geography, etc., would in one way or the other be very useful. No less important would be a wholesome study of pre-colonial polity formations in different parts of Arunachal Pradesh and its neighbouring areas.

The development of trade and British sovereignty in north-eastern India and in the neighbouring areas of Burma, China and Tibet during the colonial period brought significant changes in the economic and political formations that existed in pre-colonial times. These changes are seen to have mostly favoured the British. Studies reveal that in other parts of India, pre-colonial urban centres declined in the 19th century. Subsequently, when peace returned, a new urban entity emerged mainly in the form of politico-administrative centres, a phenomenon also noticed in Arunachal Pradesh. These new centres overshadowed the pre-colonial trade centres. Revival of trade fairs in Assam was attempted only with a mercantile motive by the East India Company, as is reflected in its report of 1797, which states: 'creating a demand for articles, the products of Europe and Bengal and aimed at getting supplied with valuable returns' (Acharya 1983, 30). The quest for profit is clear from Notes by David Scott on Welsh's Report on Assam. It is stated:

> The quality of goods of European manufacture has of late increased, and it is probable that there may be a considerable outlet for woollens at an annual fair held on the confines of Darrung, to which merchants from Thibet [sic] and the intermediate country [Arunachal Pradesh] resort. (Mackenzie 1979, 387)

The changes in the trade policy of the Tibetans too had an effect on the trade interests of Arunachal Pradesh. There was a significant change in the exchange rate of the *tanka*. It is reported that the Tibetans had raised the exchange value of their *tanka* in terms of the Indian rupee. Previously, the ratio was that of 8 *tankas* to 1 rupee, which was changed to 4 *tankas* to 1 rupee.[16] The economic fortunes of the tribes of Arunachal Pradesh declined even further when cross-border trade between India and China almost came to a standstill in the aftermath of the 1962 war.

Thus, based on the above discussion, it can be argued that extensive and efficient trading ties existed between Arunachal Pradesh and Tibet in the pre-colonial period. These ties were marked by exchanges of various commodities and items, including food, clothing, handmade paper, salt, colours and dyes and livestock. The colonial British power rather rudely interrupted these trading linkages through its profit-making activities. Therefore, perhaps the post-colonial Indian state would do well to study these pre-colonial practices to see if they can provide any guidance to contemporary policymaking, especially to the Act East policy.

NOTES

1. File No. *Tr. 54/1945 Assam Administration 'A'*.
2. Cited in Thakur (2005).
3. S. Gupta, *British Relations with Bhutan* (20) as cited in Barpujari (1994, 125).
4. For more information on Hajo, see Nath (1978, 37), Barpujari (1990, 254–255) and Mukherjee (1992, 29).

5. 'A temple complex called Viswakarma *Than* near the Shingri Hills at Dhekiajuli in the district of Darrang in Assam is visited by the Sherdukpens and Monpas of Dirrang area. The name of this complex arose from a legend according to which it was built by the god named Visvakarma' (see Sarkar 2006, 91–94).

6. See Letter No. BPG/DEV/B/10/98 dated 24.8.1998 from the E.A.C. Bhalukpong to the Director of Research, Itanagar.

7. See, Confidential Letter No. 76/63, dated 7 September, 1973 on the subject 'Emancipation of Slaves in Arunachal Pradesh'. The letter is from DC, Siang District to the Deputy Director of Research, Arunachal Pradesh Administration, Shillong, and is housed in the State Archives, Government of Arunachal Pradesh, Itanagar.

8. Outcome of field studies conducted by the author.

9. Outcome of field studies conducted by the author.

10. See file No. 2077, 'Report on Tribute, other than Monastic in the Sela Sub Agency', 1946, in the State Archives, Government of Arunachal Pradesh, Itanagar.

11. Outcome of field studies conducted by the author.

12. Outcome of field studies conducted by the author at Boot (now known as Jirigaon) and Khoina.

13. File No. 1903/1940 (confidential) Tour Diary of Political Officer, Balipara Frontier Tract kept at the State Archives, Government of Arunachal Pradesh, Itanagar.

14. File No. 1903/1940 (confidential) Tour Diary of Political Officer, Balipara Frontier Tract, kept at the State Archives, Government of Arunachal Pradesh, Itanagar.

15. Outcome of field studies conducted by the author.

16. See Tribal—A, Nos. 1–41, March 1941 in the State Archives, Government of Arunachal Pradesh, Itanagar.

REFERENCES

Acharya, N. N. 1972. 'The Trade Routes and Means of Transport in Ancient India with Special Reference to Assam'. In *Early Indian Trade and Industry*, edited by D. C. Sircar, I(viii-A). Calcutta: University of Calcutta (Lectures and Seminars).

———. 1983. *Historical Documents of Assam and Neighboring States*. New Delhi: Omsons Publications.

Barpujari, H. K. ed. 1990. *The Comprehensive History of Assam (Volume I)*. Guwahati: Publication Board of Assam.

———. ed. 1994. *The Comprehensive History of Assam (Volume III)*. Guwahati: Publication Board of Assam.

Barua, K. L. 1966. *Early History of Kamrupa*. Guwahati: Lawyers' Book Stall.

Bhattacharya, N. N. 1995. *Religious Culture of North-Eastern India*. New Delhi: Manohar.

Choudhury, P. C. 1987. *The History of the People of Assam to the Twelfth Century A.D.* Guwahati: Spectrum Publications.

———. 1988. *Assam–Bengal Relations*. Guwahati: Spectrum Publications.

Cosh, J. M. 1837. *Topography of Assam*. Calcutta: Bengal Military Operation Press.

Hamilton, F. 1940. *An Account of Assam*, edited by S. K. Bhuyan. Guwahati: DHAS.

Mackenzie, A. 1979. *The North East Frontier of India*. New Delhi: Mittal.

Mukherjee, B. N. 1992. *External Trade of Early North Eastern India*. New Delhi: Har Ananda Publication.

Nath, Rajmohan. 1978. *Background of Assamese Culture*. Guwahati: Dutta Barua.

Neog, Maheshwar. 2008. *Religions of the North-East*. Guwahati: Publication Board of Assam.

Sarkar, Niranjan. 2006. *Buddhism Among the Monpas and Sherdukpens*. Itanagar: Directorate of Research, Government of Arunachal Pradesh.

Sarma, D. ed. 1981. *KamrupaSasanavli*. Guwahati: Publication Board of Assam.

Thakur, Amrendra Kumar. 2005. 'Socio-Economic Formations in Pre-Colonial Arunachal Pradesh: Myth and Reality'. *The Indian Historical Review* 32(2):37–63.

———. 2012. 'Technology-Aided Trade in Precolonial Arunachal Pradesh'. *Resarun* 36:7–14.

———. 2014. 'State, Society, and Trade: Issues in the Historiography of Northeast India'. *Proceedings of North East India History Association*, 35th Session: 12–25.

———. 2016. *Pre-Colonial Arunachal Pradesh: Society, Technology and Trade*. New Delhi: Akansha Publishing House.

Chapter 9

Reviving Border Trade and Tourism along Nathu La in Sikkim

Dechen Bhutia

INTRODUCTION

One way of defining borders in international relations is to understand the limits of a sovereign state's space and territory. According to Salter, 'they represent the limit of the state, the space of law, authority, and responsibility' (Salter 2011, 515). Malcolm Anderson, who uses the term 'frontiers' instead of 'borders', states that we as human beings have always had a 'sense of territory' since time immemorial. The purpose of frontiers, for states, was 'to establish absolute physical control over a finite area and to exercise legal, administrative and social control over its inhabitants' (Anderson 1996, 189).

The overarching control and command of the state over its borders and hence its citizens has been one of the most important functions of modern nation states. Spatial and territorial imagination is central to the conceptualization of borders between states, and the divide—real or imagined—creates spaces for immediate borderlands and border communities. Though borderlands are located relatively far away from the major centres of polity and economy, they are not areas of stasis as often conceived but can be sites of cross-cultural communication and trans-border economic exchanges and spaces for initiating cooperation and better understanding between two nation states. By taking the case of border trade and tourism along Nathu La in Sikkim, this chapter seeks to underline the significance of positive ties between border nations and the benefits accrued to border communities. For this purpose, I rely on both primary and secondary sources, news reports, interviews and fieldwork data collected during my study on the Kalimpong–Lhasa trade route in 2016–2017.

SIKKIM FRONTIER AND THE ANGLO-TIBETAN WOOL TRADE IN THE 19TH CENTURY

In the Himalayan region, trade along the various passes has been a traditional phenomenon since ancient times. Despite the rough terrain and harsh weather, cultural contacts between various Himalayan kingdoms and their peoples shaped the trade and politics of this region. Trade missions,[1] from the various Himalayan kingdoms in the Indian subcontinent to Tibet, and the annual pilgrimages of the Tibetans to the holy Buddhist sites in India and Nepal had been a regular feature for centuries.

In the 19th century, Sikkim was a small Buddhist kingdom in the Eastern Himalayas under the rule of the Namgyal dynasty that had ascended to the throne in 1642 after its pact with the Lepchas (Risley 1894). The cultural and religious affinities of the Chogyals and the aristocracy in Tibet had for a long time defined the frontier relations between Sikkim and Tibet in the early centuries of their rule. The 'patron–client' relationship (*choyon*)[2] that defined Sino-Tibetan relations did not have a direct influence on Sikkim, as the latter enjoyed considerable autonomy and was relatively far away from the political centres in Tibet and the Chinese empire. Unlike Tibet, Sikkim was constantly preoccupied with the Gurkha invasions from the west. Subsequently, the British East India Company arrived on the scene and changed the geopolitics in the Himalayan region.

The East India Company was determined to find an alternative route to the Chinese mainland, as the one via the canton was fraught with many difficulties. The company was optimistic that a land route via the Himalayas through Tibet would make a good alternative. Several missions were then sent to Tibet, starting from George Bogle, Hamilton, Turner and Manning (Cammann 1951). Not all of them were successful, as Tibet by then had strictly held on to its 'closed-door policy', earning the epithet of 'the Forbidden Kingdom' in Europe's orientalist imaginings, north of the Himalayas.

It was only after 1861, when Sikkim became a protectorate kingdom under the British Empire, that the extent of European geopolitics and mercantilist considerations in this part of the subcontinent was unravelled. By this time, the industrial revolution in Britain had already entered its mature phase. The search for new markets for raw materials and finished goods in Asia reveals a lot about the underlying imperatives for colonization and territorial expansions.

Wool had been Tibet's most important item of export for centuries. Due to its fine quality, the British began to import considerable amounts of wool from Tibet for its factories in Manchester and London by the end of the 19th century. In fact, after Younghusband's expedition to Lhasa in 1905, the wool trade steadily increased, as new trading marts along the Sikkim frontier were established, and this is how the famous 'silk and wool route',[3] a rough pony track through the Jelep La connecting Kalimpong in North Bengal to Lhasa in Tibet, emerged as one of the major trans-Himalayan trade routes by the early 20th century. By the 1940s, 'the export of wool from Tibet amounted to 106,615 maunds (approximately 8,724,430 pounds), at 55 rupees per maund' (Harris 2008, 207).

Following India's independence in 1947, Sikkim became a protectorate of India, and after Tibet's unsuccessful bid for its independence from Communist China, Tibet was fully incorporated by the latter in 1951. Thousands of Tibetan refugees fled into the Indian subcontinent, and India was faced with the reality of a direct frontier with Communist China. For centuries, Tibet had been a natural buffer between India and China. For both these new republics, the inevitability of a long and undefined frontier along the Himalayas became the cause for border disputes. Interestingly, India was the first country to recognize the

People's Republic of China and sign the 'Agreement on Trade and Intercourse between Tibet region of China and India', also popularly known as the Panchsheel Agreement. The agreement mentioned the passes that could be open for border trade between India and China.[4] However, as relations between the two countries eventually turned sour, a border war broke out in 1962, undoing the good intentions of the Panchsheel Agreement so carefully crafted 9 years earlier. It was thereafter that India decided to close down all the border trade passes to China and stepped up its defences along the India–China border.

IMPACT OF CLOSURE OF THE TRADE PASSES ON THE BORDER COMMUNITIES

The closure of the trade passes for border communities meant the crippling of their local economy and doing away with employment opportunities that were centred on this cross-border trade. For instance, Kalimpong, which was a major commercial centre and also a bustling entrepôt along the wool and silk route in the 19th and early 20th centuries, became a dry town when the wool stopped flowing and the godowns or warehouses had to be shut down. Credit and banking facilities that financed this cross-border trade also came to a halt. Many had to seek alternative employment sources in the plains, and migration to the cities became a common phenomenon. As for the route itself, a once-bustling pony track was now hidden and covered by the thick wilderness.

OPENING OF NATHU LA FOR BORDER TRADE

In July 2006, after nearly four decades, the once-closed 'old silk route'—Nathu La—in Sikkim was opened for border trade.[5] The event was considered historic and symbolic in many ways besides the fact that the route had been reopened. It was the outcome of negotiations at various levels. In 2003, during Prime Minister Atal Bihari Vajpayee's visit to China, the possibility of restoring the former trade route was mooted, wherein both the countries signed a Memorandum of Understanding (MoU) on expanding border trade.[6]

There were also demands by the local traders in Sikkim to open the erstwhile trade route. In 2005, the then chief minister of Sikkim, Pawan Chamling, set up the Nathu La Trade Study Group. The group, led by Dr Mahendra P. Lama, the then chief economic adviser to the chief minister of Sikkim, came up with a 300-page document after consultation with a wide array of stakeholders. The group had forecasted that trade through Nathu La would top ₹2,200 crore within 5 years of its opening (Nathu La Trade Study Group 2005). However, even after a decade has passed, trade is negligible, with an annual turnover of less than ₹10 crore per year. Hence, doubts over the promised trade through Nathu La remain for most traders and analysts. However, according to Singh (2009),

> since the disputed India-China border remains the most intractable problem, it has been contended that the border trade must be viewed in terms, not of its statistical value, but its value as an important confidence-building measure, leading to transformation in the lives and perceptions of people living in the remote border regions of India and China.

Hence, the border trade between India and China could become a significant confidence building measure and set the stage for future border negotiations.

For India, the border trade agreement between India and China along Nathu La is significant for two reasons. First, it aims to revive the historical frontier trade that took place across the Sikkim–Tibet frontier, though in a completely new context and circumstances. Second, China's earlier stance of Sikkim being a contentious territory would be put to rest through engaging in border trade with Sikkim, the 22nd state of the Indian Union, as it would amount to China recognizing Sikkim as part of India. Article 1 of the MoU[7] clearly mentions, 'The Indian side agrees to designate Changgu of Sikkim state as the venue of border trade market; the Chinese side agrees to designate Renqinggang of the Tibet Autonomous Region as the venue for border trade market'.[8] As a quid pro quo, India also 'recognized' Tibet as part of China.[9]

Nathu La attracted more public attention as it was near Gangtok, the capital of Sikkim. However, its symbolic opening encouraged the people of Kalimpong and Darjeeling in West Bengal to demand the opening of Jelep La, which was linked in a way to the historic 'old silk route' and was more convenient due to its all-weather access. As the current trade through Nathu La is exclusively reserved for the residents of Sikkim, the demand for opening Jelep La gained strength.

INDIA-CHINA BORDER TRADE AT NATHU LA

The border trade through Nathu La takes place every year for roughly 7 months (usually from May to November) when the weather is more favourable and unhindered by snowfall. At present, only the people of Sikkim are permitted to trade after obtaining a trade pass granted by the state government. On either side of the border are two designated trade marts up to which the traders from both sides have access— Sherathang on the Indian side and Renqinggang (Tibetan Autonomous Region) on the Chinese side. At present, India allows import of only 15 items and export of 29 items. Some articles of import are goat and sheep skin, horses, raw silk, wool, yak hair and tails, china clay, borax, butter, common salt and szaibelyite. Indian exports through Nathu La comprise agricultural implements, blankets, copper products, clothes, textiles, cycles, coffee, tea, rice, vegetables, tobacco, spices, shoes, utensils and watches.

As seen above, the trade basket is limited in scope and has similarity with the trade basket of four decades ago. Moreover, there is a ceiling on the value of import or export for each trader, that is, ₹1 lakh per day. Besides, the trade timings are also reported to be problematic for the traders—4 days a week from 7:30 a.m. to 3:30 p.m. (Indian Standard Time) and 10 a.m. to 6 p.m. (Chinese time)—and much of the time is spent travelling through 'poor road conditions' from Gangtok and completing the formalities, with little time to trade (Elumalai 2009). Table 9.1 shows the value of the India–China cross-border trade through Nathu La.

It is interesting to note that even though China has a huge trade surplus in the overall trade balance with India (bilateral trade, including border trade), a micro-study of border trade statistics and more specifically the India–China Nathu La border trade reveals far greater exports to China through the border (see Table 9.1). Besides, it points to the overall unique characteristics of border economies vis-à-vis the

Table 9.1 *Value of India's Import and Export through the India–China Border Trade at Nathu La*

Year	Import	Export	Total Trade
2013–2014	11,670,489	77,526,935	89,197,425
2014–2015	123,620,000	160,443,830	284,063,830
2015–2016	110,486,636	602,569,967	713,056,603
Total	245,777,125	840,540,732	1,086,317,857

Source: Government of India, Ministry of Commerce and Industry, Department of Commerce (DoR/CBEC). https://commerce.gov.in/writereaddata/UploadedFile/MOC_636062753423971544_LS20160808.pdf

discourse of mainstream economy often centred on the major financial centres of a country. Though border trade constitutes a small portion of a country's bilateral trade, its importance cannot be undermined from the perspective of border regions.

BORDER TOURISM ALONG NATHU LA

Divided into four districts—East Sikkim, West Sikkim, North Sikkim and South Sikkim—the state of Sikkim has a varied topography and climate. From alpine mountains and valleys to dense forests, rivers, spectacular waterfalls and diverse flora and fauna, the state is designated as part of the global biodiversity hotspot (Arrawatia and Tambe 2011). This chapter focuses on two districts, North Sikkim and East Sikkim, as they both lie on the India–China border. Nathu La is located in East Sikkim, and this district also borders West Bengal and Bhutan. Doko La, straddling the Doklam tri-junction, also lies adjacent to East Sikkim.

In the course of my fieldwork on the old silk route along Jelep La and Nathu La, one observation was the striking beauty of the landscape and sparsely populated border towns and hamlets. Though in India most of the border regions are also designated as restricted areas, the system of an 'inner line permit' issued by the state government authorities does allow movement and passage for tourists. Table 9.2 shows the inflow of tourists over the period 2011–2017.

Table 9.2 *Tourist Arrivals in Sikkim (2011–2017)*

Year	2011	2012	2013	2014	2015	2016	2017
Domestic tourists	552,453	558,538	576,749	562,418	705,023	740,763	1,375,854
Foreign tourists	23,945	26,489	31,698	49,175	38,479	66,012	49,111

Source: Compiled from 'Statistics of tourist arrivals in the state of Sikkim', Tourism and Civil Aviation Department, Government of Sikkim. http://sikkimtourism.gov.in/Webforms/General/DepartmentStakeholders/TouristArrivalStats.aspx

The tourism industry is a major source of employment for the hill people and revenue for the state. Ensconced amid the majestic views of Mount Kanchenjunga—the world's third highest peak—pristine lakes, rich flora and fauna, temples, monasteries and the rich cultural heritage and hospitality of the people, Sikkim has immense potential for tourism and related activities. In recent years, the inflow of tourists—both domestic and foreign—has increased tremendously. Nathu La was also opened as an alternative and shorter route for the Kailash–Mansarovar yatra pilgrims on 22 June 2015, after Xi Jinping's visit to India to improve Sino-Indian bilateral relations.[10]

Some iconic tourist points in North Sikkim and East Sikkim are Changu Lake, Nathu La, Lachen, Lachung, Yumthang and Gurudongmar Lake. An influx of tourists means the demand for more hotel and lodging facilities, besides a whole lot of allied services—transport, tour guides, porters and restaurant business. Taking a stroll through the capital, Gangtok, one can see various signboards and hotels in every nook and corner of the city. However, with the popularity of 'off-the-track' adventure and ecotourism in the hills, the number of homestays in remote villages and towns has considerably increased.

In one of the field interviews with an owner of a homestay in Zuluk, a border village near Nathu La and Jelep La, he mentioned the changes in the local village economy and how earnings from tourism were now their major source of income, compared with their forefathers who depended solely on agriculture.[11] The concept of 'homestays' is a novel approach in the tourism industry and quite popular in recent times. It can be defined as 'a form of tourism that develops micro-enterprise and employment through household-owned and operated accommodation, and through related-guide services and interpretation that would enhance a visitor's experience of villages and their surroundings' (Chettri et al. 2008, 128). Homestays also provide a unique opportunity for the tourists to understand the local cultures and traditions through the 'use of décor, cuisine, and buildings, while encouraging cultural and environmental conservation' (Chettri et al. 2008).

Hence, for most of the border inhabitants in North Sikkim and East Sikkim, the once-idyllic hamlets are today dotted with numerous homestays, hotels and lodges and have played an important role in employing these communities living in such remote areas, alleviating poverty and developing the communities. Moreover, it has been noticed that from April–May to June–July, when the tourist season is at its peak and Nathu La is open for border trade, visiting the trade marts and 'experiencing' border trade also becomes part of the overall tourist experience. Thus, uniquely, border trade and tourism often complement each other.

BORDER SECURITY AND BORDER COMMUNITIES

In June 2017, a major stand-off took place between India and China along the Sikkim–Tibet border adjoining the Doklam plateau, triggered by the Chinese road-building activity in the disputed Doklam plateau between India, Bhutan and China.[12] The annual border trade across Nathu La and the Mansarovar pilgrimage were also disrupted because of this flare-up between the two nations. Even though the road that China was building on the Doklam plateau might have been only a dirt road, it would not be seen as a minor development through the lens of security due to two major reasons—first, the nature of this territory, as India and Bhutan claim it to be disputed, and second and more importantly, its proximity to the Siliguri Corridor, a 27-km-long stretch of the land corridor (also known as the Chicken's Neck) that connects mainland India to its North Eastern Region.

Because of the geo-strategic location of this region and its proximity to the Siliguri Corridor, which is the only vital link to the North Eastern Region of India, any instance of border tension between the two giant Asian neighbours has the potential to alter normalcy and peace in this sub-region. Conflict in any form has the potential to disrupt any efforts towards mutual understanding and confidence building across the borders. Furthermore, it would also disrupt traditional cross-border economic and cultural exchanges, thus affecting the people's livelihoods.

The Sikkim–Tibet trade route, also known as the 'old silk route', which was once the most important and busiest of all Indo-Tibetan trade routes, had a promising start during the early 20th century until the 1962 Sino-Indian War. The remains of this route also lie buried in the history of this sub-region. Though at present it might be a small border outpost along the India–China border, it has not lost its significance.

CONCLUSION

Today, India and China are two of the largest economies in the world and also represent Asia's dawn in global politics. As India's Act East policy aims to foster better relations with Southeast Asian countries and as China embarks on its One Belt One Road initiative, a common thread runs across both these initiatives, that is, building bridges across borders and linking various countries for peace and prosperity. India and China are geographically contiguous in the Asian landmass, and hence peaceful coexistence is the only way for the survival of both nations. Historically, before borders were drawn as lines on maps, communities on either side of borders had sociocultural and economic linkages. Understanding borders also entails understanding the history of border communities. Today, borders are studied more from a security perspective, which in a way also reflects how meanings of space and place change over time, which is amply reflected in the India–China border. Nonetheless, when countries cooperate despite differences, meaningful changes are often translated to other spheres. In the case of the border communities in Sikkim, reopening of the border trade through Nathu La and the fast-growing tourism industry have positively influenced the lives of border communities and, if the trend continues, would do so in the future.

NOTES

1. For instance, Ladakh's 'lo phyag' mission from Leh to Lhasa and Nepal's tribute mission to China through Tibet. For more details, refer to Bray and Gonkatsang (2009). Furthermore, Buddhism travelled from Nepal to Tibet, where later a unique form of Buddhism called 'Tibetan Buddhism' evolved and is the major form of Buddhism in the Himalayan region till today.
2. The Qing dynasty, also called the Manchu dynasty, established its protectorate over Tibet based on the priest–patron relationship—*choyon*—and served as the patron of Tibetan Buddhism.

3. Wool was an important commodity traded along this route. Hence, instead of the popular term 'Silk Route', the article uses the term 'silk and wool route'. For more details, refer to Harris (2008).
4. 'Agreement Between the Republic of India and The People's Republic of China on Trade and Intercourse between Tibet Region of China and India', Peking, 29 April 1954. For details, see Media Center (1954).
5. In June 2003, a Memorandum of Understanding was signed between the government of the Republic of India and the government of the People's Republic of China (PRC) on expanding border trade.
6. For more details, see Ministry of External Affairs (2003).
7. Ibid.
8. Ibid.
9. This coincided with Vajpayee's visit to China in 2003. For more details, refer to Parthasarathy (2003).
10. In 2015, China opened the Nathu La as an alternative route for the annual Kailash–Mansarovar pilgrimage. For more details, see *Livemint* (2015).
11. Personal interview with G. Pradhan, 2017.
12. There were reports that China was building a road near Doklam, a strategic area in the Eastern Himalayas bordering India, Bhutan and China. For more details, see *Pioneer* (2017).

REFERENCES

Anderson, Malcolm. 1996. *Frontiers: Territory and States Formation in the Modern World*. Malden, MA: Polity Press.

Arrawatia, M. L., and S. Tambe. eds. 2011. *Biodiversity of Sikkim: Exploring and Conserving a Global Hotspot*. Gangtok: Information and Public Relations Department, Government of Sikkim.

Bray, John, and T. Gonkatsang. 2009. 'Three 19th Century Documents from Tibet and the Lo Phyag Mission from Leh to Lhasa'. *RivistadeglistudiorientaliSupplemnto 2: Impagninato*, 17(28):97–116.

Cammann, Schuyler. 1951. *Trade through the Himalayas: The Early British Attempts to Open Tibet*. New Jersey: Princeton University Press.

Chettri, N., E. Kruk, and R. Lepcha 2008. 'Ecotourism Development in Kanchenjunga Landscape: Potentials and Challenges'. In *Biodiversity Conservation in the Kanchenjunga Landscape*, edited by Nakul Chettri, Bandana Shakya, and Eklabya Sharma, 123–132. Kathmandu: International Center for Integrated Mountain Development (ICIMOD).

Elumalai, K. 2009. 'Indo-China Border trade through Nathula: Prospects for Growth and Investment in Sikkim'. In *Sikkim's tryst with Nathu La: What Awaits India's East and Northeast?* edited by J. K. Ray and R. Bhattacharya, New Delhi: Anshah Publishing House.

Harris, Tina. 2008. 'Silk Roads and Wool Routes: Contemporary Geographies of Trade Between Lhasa and Kalimpong'. *India Review*, 7(3):200–222.

Livemint. 2015. China opens Nathu La pass for Kailash Mansarovar pilgrims. *Livemint*, 23 June. http://www.livemint.com/Multimedia/ljhs7LoIB-3BZQvrwQSigCP/China-opens-Nathu-La-pass-for-Kailash-Mansarovar-pilgrims.html?gid=1&utm_source=scroll&utm_medium=referral&utm_campaign=scroll (accessed 8 March 2018).

Media Center. 1954. *Agreement on Trade and Intercourse with Tibet Region*, 29 April. Ministry of External Affairs, Government of India. https://www.mea.gov.in/bilateral-documents.htm?dtl/7807/Agreement+on+Trade+and+Intercourse+with+Tibet+Region (accessed 10 March 2020).

Nathu La Trade Study Group. 2005. *Sikkim-Tibet Trade via Nathu La: A Policy Study on Prospect, Opportunities and Requisite Preparedness.* Report submitted to the Government of Sikkim.

Parthasarathy, G. 2003. 'Vajpayee Visit- Foreign Policy lessons from China'. *Business Line*, 3 March. https://www.thehindubusinessline.com/2003/07/18/stories/2003071800030800.htm (accessed 8 March 2018).

Pioneer. 2017. China building Road near Doklam for year-round logistical support to PLA. *Pioneer,* 13 December. http://www.dailypioneer.com/todays-newspaper/china-building-road-near-doklam-for-year-round-logistical-support-to-pla.html (accessed 6 March 2018).

Risley, H. H. 1894. *The Gazetteer of Sikhim*. Delhi: Oriental Publishers (reprinted 1973).

Salter, M. B. 2011. 'Border, passports, and the global mobility'. In *The Routledge International Handbook of Globalization Studies,* edited by B. S. Turner, 514–530. London: Routledge.

Singh, Swaran. 2009. 'Limitations of India- China Economic Engagement'. *China Report*, 45(5):285–299.

Chapter 10

Tripura–Bangladesh Borderlands
Socio-economic Significance of Border *Haats*

Suparna Bhattacharjee

INTRODUCTION

Haat, a local name familiar to the people in the Indian subcontinent, refers to simple informal markets that primarily function on fixed days of the week within a set time frame. A border *haat* is a makeshift market set up in some location in a border area to facilitate exchange of goods and items among people across the border. It is an open-air cross-border market, free from the formal arrangements and standard regulations associated with formal cross-border trade. Such border markets link producers and consumers belonging to different countries and even provide an opportunity to small traders to market their products across the border, which otherwise would not have been feasible. Such commercial openings are arranged primarily to offer a platform to small local entrepreneurs to sell their products to local consumers and enable them to enjoy the benefit of getting their daily requirements at an affordable price. This chapter looks into the functioning of two border *haats* in Tripura along the Indo-Bangla international border to understand the socio-economic dimension of *haat*s and their significance in the life of the people living in the border areas between India and Bangladesh. It is based on an empirical study of Kamalasagar–Tarapur *haat* and Srinagar–Chagalnaiya *haat* in the state of Tripura.

GLOBAL EXPERIENCE

Instances of such informal cross-border trade markets are evident in various parts of Asia and Africa. The Pasar Serikin border market was opened in 1992 on the borders of Indonesia and Malaysia. The market opens every Saturday and Sunday in Serikin[1]—bordering Sarawak province of Malaysia and Kalimantan town of Indonesia. It is a localized market that caters mainly to the needs of the people of Sarawak. The products like household items ranging from kitchen utensils and curtains to Islamic prayer items, handicrafts and accessories are sold from Kalimantan. Apart from the locals, people from adjacent parts of Singapore, Indonesia and Malaysia come to the market as consumers. The market is jointly operated by the locals of Kalimantan from across the border and people from Sarawak. According to a study, on market days, each trader earns around US$15–20, with a sizeable profit margin (Awang et al. 2013).

A significant proportion of such informal regional cross-border trade has been in vogue, in both the sub-Saharan region and the Great Lakes region of Africa. For families living in the border regions of Burundi, Rwanda, Uganda and some other states of the Great Lakes region, border markets become a source of livelihood. The participant vendors here are predominantly women. Mostly local agricultural produce, such as cereals, pulses, vegetables and fruits, are traded. According to the World Bank and European Economic Commission, informal border trade accounts for a significant portion of the Great Lakes region's economy (Breton and Isik 2018). The share of informal border trade represents 25 per cent of Rwanda's total trade statistics, whereas it accounts for 55 per cent of Uganda's trade. It may be noted here that Rwanda and Uganda have joined COMESA (Common Market for Eastern and Southern Africa, which was formed in 1994) to set up trade facilitation centres to encourage such small-scale commercial joints. Informal border trade is a significant source of livelihood for people living in the fringes of the Great Lakes region of Africa. In sub-Saharan Africa (the part of Africa lying south of Sahara, which includes countries like Zambia, Congo, etc.), one comes across informal cross-border markets. Informal cross-border trade contributes almost 40 per cent to the gross domestic product (GDP) of each nation (Lesser and Leemam-Moise 2009, 44).

EXPERIENCE FROM INDIA'S NORTHEAST

India's North Eastern Region (NER) has a rich tradition of *haats*. This form of markets was common between hills and plains and existed along the foothills of Khasi–Jaintia and Garo hills. Recorded evidence of such *haats* has been available in literature since the 13th century. *Haats* used to take place at regular intervals—usually on a fixed day of the week—between the erstwhile Himas (traditional polities of the indigenous communities in Khasi–Jaintia Hills of present day Meghalaya) and the plain areas of erstwhile East Bengal (present-day Bangladesh) and the Ahom state (present-day Assam). The produce such as paan leaves, bay leaves, honey cotton, lime and potatoes from the hills were in great demand in the plains, whereas the items such as rice, fish, salt, spices, copper and silk, which were produced in the plains but were not available to the people of the hills, were available for exchange in the *haats*. The *haats* offered a platform for production and exchange of goods based on the comparative advantage of soil/climatic condition of the respective areas to fulfil the mutual needs of the people. They were a source of revenue for the Himas. In the beginning, the barter system was the mode of exchange, but gradually the commercial activities took place in terms of monetary exchange. *Haats* reached their pinnacle during the British administration in the region. With the consolidation of British administration, a communication network developed in the region. Construction of roads, brittle paths, bridges, etc., helped these *haats* become a thriving hub of commercial activities. Many British historians and administrators have documented the profitable commercial transaction at *haats*. On some items, transactions worth lakhs of rupees were recorded in the early part of the 19th century. However, the *haats* came to a closure with the partition of the subcontinent. After a brief period of being functional in the year 1965,[2] the *haats* were closed again. This brought untold sufferings to people living in the border areas, depriving them of livelihood and access to many essential goods.

REOPENING OF *HAATS* IN INDIA'S NORTH EASTERN REGION

The prospect of reopening of *haats* became a possibility during the 3-day state visit of the then Prime Minister of Bangladesh, Sheikh Hasina, to India on 10–13 January 2010. The opening of border *haats* was envisaged to generate livelihood for the people of border areas. Bangladesh and India agreed to a 'Comprehensive Framework of Co-operation for Development' encapsulating the mutual vision of future cooperation in the fields of water resources, transportation, connectivity, education and tourism. The agreement highlighted the desire of the two nations to revive the traditional links and launch a new phase in the bilateral relations. The ministry of commerce of the two countries took the lead in finalizing a Memorandum of Understanding (MoU) to fulfil the goal of renewal of traditional links and the shared past (Ministry of External Affairs 2010). Within 8 months of the publication of the joint communiqué, the two countries signed an MoU on 23 October 2010, leading to the opening of border *haats* at the designated places along the Indo-Bangla border. The MoU holds immense significance in the sense that it was a follow-up to the decision taken at the prime ministerial–level meeting between the two countries. The terms such as 'shared history', 'traditional links' and 'connectivity' mentioned in the communiqué got contextualized in the form of the MoU. The very first paragraph of the MoU referred to the practice of the traditional system of marketing (*haat*) in the region.

Recognizing that 'border *haats*' promote the well-being of the people dwelling in remote areas across the borders of the two countries, through establishing a traditional system of marketing the local produce through local markets, and recalling the decision of the then prime minister of Bangladesh H. E. Sheikh Hasina and Manmohan Singh, the then prime minister of India, during the former's visit to India on 10–13 January 2010, both the countries have agreed to establish *haats* at selected places, including Meghalaya: '*Haats* shall be established on pilot basis in selected areas, including Meghalaya border to promote the wellbeing of the people dwelling in remote areas across the borders of the two countries'.[3] Subsequently, the two governments entered into another agreement identifying new border *haats* (PMINDIA 2016).

BORDER *HAATS* IN TRIPURA

The border *haats* first materialized along the Meghalaya–Bangladesh borders. Tripura became the second state in the NER to set up such *haats* along the Indo–Bangla international border. Exhibiting a keen interest in establishing *haats*, the government of Tripura proposed establishing eight border *haats* to the Ministry of Commerce and Industry in eight specified locations along its long border with Bangladesh. Tripura is surrounded by Bangladesh on its north, south and west and shares an 856-km-long border with Bangladesh, which forms 84 per cent of its total land border. The Ministry of Commerce and Industry, however, sanctioned two *haats*. A second MoU was signed between Bangladesh and India in 2014, comprising 15 articles to this effect. The MoU documents the various provisions related to the functioning of *haats* in Tripura. It was decided to set up two *haats* as pilot projects to evaluate their feasibility, establishment and operational costs and unforeseen adverse effects and to improve upon the *haats'* future designs.

The *haats* are located at the zero line on the international border between India and Bangladesh. Kamalasagar is the border town of Sipahijala district in South Tripura, which shares a border with Tarapur town of Brahmanbaria district of Bangladesh. The Srinagar–Chagalnaiya *haat* is in Pohang Bari block of Sabroom subdivision of South Tripura district, which shares a border with Purba Madhugram of Chagalnaiya subdivision of Feni district, Bangladesh. The Srinagar–Chagalnaiya border *haat* was inaugurated on 13 January 2015 by the then Indian Minister of State (Independent Charge) for Commerce and Industry, Mrs Nirmala Sitharaman, and her Bangladeshi counterpart, Mr Tofail Ahmed. The *haat* is at Border pillar No. 2195/6 on the zero line at the Srinagar–Chagalnaiya border. The Kamalasagar–Tarapur *haat* is at Border Pillar No. 2039, on the zero line at the Kamalasagar–Tarapur border. The *haat* was jointly inaugurated through a video conference by Indian Prime Minister Mr Narendra Modi and his Bangladeshi counterpart Mrs Sheikh Hasina on 6 June 2015. It was opened to the public 5 days later, that is, on 11 June 2015. On the Indian side, the departments concerned with the overall supervision of the *haats* are the Ministry of Commerce and Industry, Government of India, Department of Industries and Commerce, Government of Tripura, and Tripura Industrial Development Corporation Limited (TIDC). TIDC is the implementing agency.

BORDER *HAAT*: STRUCTURE AND MODE OF OPERATION

The border *haats* operate on a fixed day of the week within stipulated hours. The *haats* start from 10 a.m. and continue till 3.30 p.m. The vendors get additional time (30–40 minutes to set up and dismantle stalls). The *haats* function for 6 hours. They function within a very modest structure that is primarily an enclosure made up of concrete walls or barbed wire. The size of a *haat* is specified as 75 m × 75 m. The layout and design have been specified in the MoU. From the pillar on the zero line, the space for a *haat* (i.e., 75 m) is measured. *Haats* have two entry/exit points—one from the Indian side and the other from Bangladeshi territory. A distance of 75 m from the pillar towards Bangladesh (up to the entry/exit point) is allocated for vendors from Bangladesh. Similarly, from the pillar to the entry/exit point towards India, vendors from India have their designated space to set up shops. Three–four long market sheds have been constructed on each side to facilitate trade. In the long shed with tin roof, space is allocated for each vendor. The location of a *haat* is decided based on the agreement between the two countries, in consultation with the Border Security Force (BSF) of India and the Border Guards Bangladesh (BGB). India bears the lion's share of the cost of construction of the *haats*. The construction of one *haat* costs the exchequer over millions. For example, the Srinagar–Chagalnaiya *haat* and the road on the Indian side were built by India at a cost of ₹24 million (₹2.44 crore) (*The Economic Times* 2015). The construction work was undertaken by TIDC, with funding from the Ministry of Commerce and Industry, Government of India.

Each of the *haats* has a Border Haat Management Committee to look after the issues related to the functioning, law, order, overall progress and prospects of the *haats*. Maintenance of the overall security measures is the responsibility of BSF and BGB. They do not enter the *haats* but strictly monitor and check the entry and exit of people. As a border area is a sensitive zone, the concern for security remains paramount. The committee includes the following officers: the additional district magistrate of the district in which the *haat* is located (in the case of India) and the district magistrate (in the case of Bangladesh), the commanding officer of the border outpost of BSF, the commander of BGB, the officer in charge of the local police stations, customs officers and representatives of panchayats (local self-governments) of both the countries.

The *haats* function within certain confines embedded in their own operation, objectives and nature. The modalities of operation are part of the MoU. The MoU is valid for 5 years. At the end of 5 years, the MoU may be renewed by the two parties for an extension of another 5 years. The MoU may also be suspended on mutual consent. Five hundred vendees/visitors from each side are allowed to enter a *haat*. Half an hour before the beginning of a *haat*, entry passes are issued by the management committee. Each *haat* has a counter on either side of the border where the entry passes are issued to the vendors/visitors. Twenty-five vendors from each side (total 50) have the permission to set up stalls. The vendors who are allowed to sell their products in the border *haats* shall be residents of an area within 10 km from the border *haat*. This regulation intends to extend an opportunity to the people of the areas closest to the border to reap the benefits of the border market. For the vendees, however, the specification that they have to be from within a 10-km radius from the border is not so rigid. In fact, in the Kamalasagar *haat*, everyone is issued a pass to enter till the stipulated number is reached. Five hundred people from either side are allowed to visit the *haat* with passes, to enter the market either as a visitor or a vendee. In the Srinagar–Chagalnaiya *haat*, the credentials of the vendees are also strictly monitored.[4] It being a smuggling-prone area, there should be strict monitoring. According to the MoU, the estimated value of purchases by an individual/family shall not be more than, in the respective local currency equivalent to, US$100. The limit may be revised through mutual consent of the *haat* management committees of the two sides.[5] In the case of the Srinagar–Chagalnaiya *haat*, the quantity of any item that a vendee can purchase is also specified, which is applicable for vendees of both sides. The list is very prominent, and it is displayed at the entry gate. In the Kamalasagar–Tarapur *haat*, such restriction was not visible. There is a strict vigil on the quantity of the goods purchased in the Srinagar–Chagalnaiya *haat*, as it is located in a sensitive and smuggling-prone area.

As regards the vendors, as mentioned earlier, 50 vendors in total (25 from either side of the border) are identified. They are familiarized with the rules and trained by the local bodies of the respective countries. The vendors are allowed to enter the enclosure under the watchful eyes of the security officials after detailed checking of their credentials. Each vendor has an identity card that is issued by the local authorities. Security remains paramount, as the *haats* are located on border areas that are considered very sensitive. On the Indian side, layers of security checks are carried out in both the locations. Only stringency varies from the Kamalasagar–Tarapur *haat* to the Srinagar–Chagalnaiya *haat* for reasons already mentioned.

The *haats* have been set up to facilitate exchange of goods primarily between the people living across the border and to cater to their daily requirements. Primarily, local and perishable goods are sold at the *haats*. In the beginning, only goods produced in areas within a 5-km radius from the border were allowed to enter the *haats* for exchange. Now, the radius has been extended to 10 km. Transactions on listed items approved by both countries take place. Sixteen types of goods and products, such as horticultural products, dry fish, minor forest products excluding timber, handlooms, wooden handmade items, etc., may be exchanged at the *haats*. The items have been listed after careful consideration of the people in the border areas and to help the communities improve their livelihood. The list, however, is not very rigid. It is negotiable. New items may be added keeping in mind their local character based on the mutual consent of the authorities of the two countries to sustain the *haats*.

In tune with the basic character of the *haats,* the traders have been exempted from the payment of custom duties. This exemption is made keeping in view the socio-economic condition of the people living in such remote border regions. The buyers purchase goods using the currency of their country. Exchange of currency is exempted from foreign exchange regulation as per the MoU. The vendors exchange currencies among themselves based on the official exchange rate (the vendors from India exchange currencies with their Bangladeshi counterparts, and vice versa). Initially, there was no bank in the *haat* areas to exchange currencies. Although banks have opened branches near the *haats*, the practice of accepting each other's currency continues at an individual level. This arrangement of a 'one-on-one' basis works with no issues.[6]

SOCIO-ECONOMIC DYNAMICS OF *HAATS*: THE ECONOMIC POTENTIAL

Haats offer limited opportunities in terms of items to be traded (16–20 items, on average), the number of vendors (50) and vendees (500) and the per-day, per-head spending cap (of US$100). Yet, their socio-economic significance cannot be overlooked. The field visits to the Kamalasagar–Tarapur border *haat* and the Srinagar–Chagalnaiya *haat* led the author to draw the inference that notwithstanding their limited spheres of operation, *haats* may prove to be a valuable mechanism to boost bilateral ties and the well-being of the communities living in border areas. *Haats* have opened a window of opportunity for the people to earn a livelihood. In the first 3 months of the opening of the Kamalasagar–Tarapur border *haat*, the Indian side had registered cash trade equivalent to 77.83 lakh in terms of Indian currency. During the same period, the Bangladeshi front recorded sales worth ₹25.63 lakh. In the next 7 months, the Indian side registered a whopping figure of ₹287.04 lakh worth of transactions. During the same period, its Bangladeshi counterpart did business worth ₹187.32 lakh, calculated in terms of Indian currency.[7] The other *haat*—the Srinagar–Chagalnaiya *haat*—also showed a similar promising trend. In the first three *haats*, the total commodity sold by the Indian vendors amounted to ₹1,892,550, and the Bangladeshi side registered 425,270 in terms of Bangladesh currency (₹361,479).[8] The transaction increased by leaps and bounds in terms of value with each successive *haat*. Garments, cosmetics, aluminium items, baby napkins (diapers) and spices are some Indian items in great demand. The items from Bangladesh which have demand from Indian vendees are melamine products, plastic items, *daos* (a kind of knife), spades, vegetables, fresh river fish, dry fish and *lungis*.

The two *haats* differ from each other in terms of demands. The Kamalasagar–Tarapur *haat* is well known for melamine crockery, fruit and sweets from Bangladesh. On the other hand, Indian garments, textiles and cosmetics are in great demand among the vendees from Bangladesh. In contrast, the Srinagar–Chagalnaiya *haat* offers a market for the Bangladeshi fish vendors and vegetable, pulse and palm jaggery sellers. Regarding the Indian vendors, the Kamalasagar–Tarapur *haat* offers a market for garments, saris, cosmetics and baby napkins (diapers). In the Srinagar–Chagalnaiya *haat*, the Indian vendors selling tea leaves, baby products, diapers, artificial jewellery, lipsticks, powders, creams and cosmetics attract several consumers. Indian cosmetic items fetch vendees from Bangladesh in both the *haats*. The obsession over Indian cosmetics is so much that one Indian visitor to the *haat* pointed out that cosmetic products with fabricated brand names also are for sale.[9] Based on a calculation from the responses of the respondents, it may be discerned that, on average, a cosmetic seller's proceeds from the business may reach ₹2–3 lakh per annum; however, it is difficult to arrive at a precise figure, as official records of profits or losses of individual vendors are not maintained.

During the field survey, it was observed that commodities worth ₹6.50 lakh were traded, on average, on every market day at the Srinagar–Chagalnaiya border market. The profit earned by each vendor on average is in the range of ₹1,500–2,000 per *haat*. The other *haat*—the Kamalasagar–Tarapur border *haat*—is also doing equally well in terms of commercial dealings. At the Kamalasagar–Tarapur *haat*, on average, commodities worth ₹8.95 lakh are sold on a *haat* day. The vendors make an average profit margin of ₹1,500–2,000 per *haat*. It was observed during the fieldwork that 75 per cent of the products brought by Indian vendors and about 40 per cent of the products brought for sale by Bangladeshi vendors were sold. It was also observed that there was an increase in the number of vendors and vendees, volume of goods sold, etc. The Sub-Divisional Magistrates (SDMs) of the Kamalasagar–Tarapur *haat* and the Srinagar–Chagalnaiya *haat* expressed their satisfaction over the performance of the *haats*.[10]

The management committees of the *haats* earn a handsome amount of money through the sale of entry passes. A guest charge of ₹20 is collected from the vendees/visitors/guests. This paltry amount, when calculated in terms of a financial year, shows evidence of a substantial total. In a year, the two border *haats* are likely to earn ₹3–4 lakh. For instance, between June 2015 and October 2016—in 16 months—the Indian side earned over ₹4 lakh—to be specific, ₹407,280—and the Bangladeshi side earned over 6 lakh in terms of Bangladesh currency, that is, BDT 603,350.[11] According to the SDM of Sabroom district, Tripura, on average, revenue from entry tickets or guest charges per *haat* amounts to ₹15,000–16,000, whereas the printing cost is hardly ₹4,000–5,000.[12] This helps in generating a source of revenue for the government—a portion of which, however, is spent on maintenance of the *haats*.

The vendors are not the only people who make a living out of the *haats*. The auxiliary service providers such as drivers running vans, buses, sumo vehicles and taxies carry goods and visitors/vendees and make a living. In response to a query, one van driver mentioned that the *haat* has come as a blessing—a source of income hitherto unlikely for many of them. Many respondents among the visitors and vendees, particularly from the Indian side, mentioned that the road condition had improved a great deal after the inauguration of the *haat* and that the frequency of buses plying up and down had also increased. The improved communication link may be counted as an 'add-on' effect of the entire project. This is more apparent in the case of the Kamalasagar–Tarapur haat.

Kamalasagar always attracts a huge crowd, as the famous Kasba Kali Mandir (temple) is just near the border on the Indian side. The area already had numerous shops satisfying various requirements of pilgrims, from shops selling puja offerings to eateries. A government guest house located there is maintained by Tripura's tourism department. Shop owners selling puja items and running food joints informed the author that their earnings had witnessed a threefold increase after the opening of the border haat. On the market/*haat* days, they earn handsome returns.

SOME FIELD OBSERVATIONS

As part of the fieldwork, several traders involved with the border *haats* were interviewed. The vendors expressed satisfaction over their earnings from the sale of items and said that the earnings helped them in addressing the family issues such as education of children, poverty alleviation and employment opportunities. The items such as sarees, local fish and melamine items are in great demand. Apart from addressing the local demands and ensuring monetary gains, the border *haats* also help in building a sense of shared identity, belongingness, reciprocity and trust among the buyers and sellers, creating a powerful bond and a network of relations. The *haats* also give opportunities to several women from either side of the border to sell and buy goods. They turn into a *mela*, meaning a local festive gathering where people from different walks of life and different political spaces meet and mingle with one another. The border *haats* provide a platform for people to meet relatives and friends living on the other side of the border and build a healthier security environment in the border region. One can buy from a wide range of products in a relaxed atmosphere. A respondent, a visitor, said,

> the *haat* is in border zone, the sight of the fenced wall, presence of security forces, was scary and a very uncomfortable sight in the beginning but now people get their daily requirements entering into an area which literally belongs to another country. This was absurd to think even few months back but now common gentry feel very happy.[13]

These words resonate with the narrative of John Agnew. Agnew is a firm believer that borders exist for a variety of purposes (Agnew 2008, 175) and that there is a need to frame border thinking away from a realist paradigm. Looking at the functioning of *haats*, one develops an impression that, notwithstanding the fact that borders are a human construct, their alienating effect may be minimized through creating a co-sphere of activities of people where 'social identities may converge and coexist' (Flynn 1997).

These positive contributions should not make one ignore the limitations. While some items are in great demand and yield profits for the vendors, some traders also complained about declining profits for some items. Some complained of the high-handed manner of the security personnel in some *haats*. The security personnel are accused of restricting the number of vendees and sometimes seizing items bought on the pretext of their being in excess of the procurement limit. It was also found that some items sold in the *haats* are bought from far-off places, sometimes even from foreign countries,[14] as they are in demand among the people on the other side of the border. It was also observed that there were middlemen who bought items from the vendees in bulk and carried them to far-off places for sale. Some have also complained of restricted hours/days of sale and lack of infrastructure and basic facilities in the border *haats*.

CONCLUSION: FUTURE PROSPECTS

Acknowledging the huge economic potential, the Ministry of Development of North Eastern Region (MDoNER) asked the central government to sanction 66 more such *haats* along Indo-Bangla border in Meghalaya, Tripura, Bengal and Assam. Further, the state governments of Arunachal Pradesh, Manipur, Nagaland and Mizoram, which together share 1,643 km of land boundary with Myanmar have also placed a demand to open *haats* along the Indo-Myanmar border. According to the MDoNER, if all the *haats* are opened, commercial transactions worth an estimated US$20 million would take place per week. Bangladesh also has expressed the desire to have more such openings. Shafikul Islam, Additional Secretary, Ministry of Commerce, Bangladesh, expressed the hope that the livelihood of people in areas adjoining the border *haats* would improve with the establishment of more border *haats*.[15] The case study shows that apart from economic benefits, the border *haats* address many other issues of concern to the people living in the borderlands. The governments need to address the teething problems that the pilot project has been facing and ensure that the border *haats* are run in more professional and humane ways, so they act as bridges between the communities separated by the accidental borders.

NOTES

1. Serikin is a small town of Malaysia on the border of Indonesia and Malaysia. The border market, apart from meeting the needs of the border communities, has become a tourist spot. For more details on Serikin border market, see SarawakVoice (2019).
2. After 1947, the eastern part of Bengal became part of Pakistan, and Government of Pakistan decided to open *haats* in 1965; however, within a few days of their opening, the *haats* were closed down unilaterally by Pakistan, which cited security concerns.
3. 'Memorandum of Understanding between The Government of The Republic of India and The Government of The People's Republic of Bangladesh',

signed on 11 January 2010 between Joint Secretaries, Department of Commerce of both Indian and Bangladesh governments.

4. The Border Security Force (BSF) takes care of the entry and exit of the Indians, and the Border Guards Bangladesh (BGB) looks after the Bangladesh side.

5. Each individual is allowed to purchase only as much of the commodities produced in Bangladesh/India as reasonable for bona fide personal/family consumption.

6. A respondent answered that after the business hours, the vendors exchange currencies among themselves based on the official exchange rate. The respondent explained that ₹100 of India is equivalent to BDT 110.

7. The figure was calculated from the official record maintained by the Office of the Sub-Divisional Magistrate, Bishalgarh: Sipahijala, Government of Tripura, Reference No. 5(74)-vol-ii/DM/SPJ/REV/2016/33795 dt 27/10/2016

8. Calculated from the records maintained by the Block Office, Poangbari, South Tripura. The record was published at the meeting of the Joint Border Haat Management Committee that took place on 24 August 2016, time 12.00IST/20.30 BST.

9. Focus group discussion with the vendees from India at the Kamalasagar–Tarapur *haat* on 14 January 2016.

10. Based on an interview with the Sub-Divisional Magistrate (SDM) of Bishalgarh on 12 June 2017 and the Block Development Officer (BDO) of Poangbari on 6 January 2016.

11. Information collected from the block office. As per the official record, the amount of money reaches up to ₹4 lakh.

12. Discussion with SDM, Srinagar.

13. Based on an interview at the Srinagar–Chagalnaiya *haat* on 24 December 2019.

14. For example, green apples from China come to Bangladesh and are sold in Tripura–Bangladesh border haats.

15. See Outlook (2018).

REFERENCES

Agnew, John. 2008. Borders on the Mind; Re-framing Border Thinking'. *Ethics and Border Politics*,1(4), Special Issue. http://doi.org/10.3402/egp.v1i4

Awang, Abd Hair, Junaenah Sulehan, Noor Rahamah Abu Bakar, Mohd Yusof Abdullah, and Ong Puay Liu. 2013. 'Informal Cross Border Trade Sarawak (Malaysia)-Kalimantan (Indonesia): A Catalyst for Border Community's Development'. *Asian Social Science*, 9(4):167–173.

Breton, Paul, and Gozde Isik. eds. 2012. *De-fragmenting Africa: Deepening Regional Integration in Goods and Services*. Washington DC: World Bank. https://open-knowledge.worldbank.org/handle/10986/12385 (accessed 29 November 2020).

Flynn, K. D. 1997. 'We are the border: Identity, exchange, and the state along the Benin-Nigeria border'. *Ethnologist* 24(2):311–30.

Lesser, Caroline, and Evdoika Leemam-Moise. 2009. *Informal Cross Border Trade and Trade Facilitation Trade Reform in Sub-Saharan Africa* (OECD Report. Paper No. 86). https://doi.org/10.1787/18166873 (accessed 5 June 2018).

Ministry of External Affairs. 2010. *Joint Communiqué issued on the occasion of the visit to India of Her Excellency, Sheikh Hasina, Prime Minister of Bangladesh*. 12 January 12. https://mea.gov.in/bilateral-documents.htm?dtl/3452/Joint+Communi qu+issued+on+the+occasion+of+the+visit+to+India+of+Her+Excellency +Sheikh+Hasina+Prime+Minister+of+Bangladesh (accessed 12 June 2017).

SarawakVoice. 2019. 'Serikin Border Market a Boon for Visitors', 12 February. https://sarawakvoice.com/2019/02/12/serikin-border-market-a-magnet-for-visitors/ (accessed 29 November 2020).

PMINDIA. 2016. '*MoU between India and Bangladesh and Mode of Operation of border haats for setting up of Border Haats, 6 January*. https://www.pmindia.gov.in/en/ news_updates/mou-between-india-and-bangladesh-and-mode-of-operation-of-border-haats-for-setting-up-of-border-haats/ (accessed 22 May 2016).

The Economic Times. 2015. 'First Border Haat Opens along the Tripura Bangladesh frontier in Tripura on January 13'. *The Economic Times*, 4 January. https:// economictimes.indiatimes.com/news/politics-and-nation/first-border-haat-opens-along-the-tripura-bangladesh-frontier-in-tripura-on-january-13/arti-cleshow/45750337.cms?from=mdr (accessed 7 Feb 2021).

Outlook. 2018. 'Six New Border Haats to be set up', 23 July. https://www. outlookindia.com/newsscroll/six-new-border-haats-to-be-set-up/1354728 (accessed 29 November 2020).

Chapter 11

Meghalaya-Bangladesh Border Trade

A Study of the Balat-Sunamganj *Haat*

Rakhal Kumar Purkayastha

INTRODUCTION

Border trade is different from trade through air, land or seaports, as trade through ports involves clearances through customs and has a large volume. Border trade, in contrast, is an overland trade that takes place in a designated area between border communities living on either side of an international border. Border trade encourages trade in certain commodities that are locally produced by the border communities. It enables the communities to sell surplus products and buy items of their needs not available in the nearby commercial areas of their own countries. Because of geographical proximity, transportation costs can be kept at a minimum in border trade.[1]

There has been a long history of border trade taking place along India's 5,182-km-long north-eastern borders. In the local dialect, the marketplaces in the villages are referred to as *haats*. It is from there that the term 'border *haat*' became popular in reference to the marketplaces in border areas where the border communities engage in buying and selling of goods. Several border *haats* were operating in the border areas. However, in recent years, efforts have been made by governments to start border *haats* in certain designated places along the international borders. Two such border *haats* started operating in the north-eastern state of Meghalaya that shares a 443-km-long border with Bangladesh. Based on a field survey undertaken by the author, the present chapter makes an effort to present the history of informal trade that existed between tribal communities and their neighbours, and it studies the actual processes at work in the Balat–Sunamganj *haat* and suggests ways and means of promoting border trade in this area.

CREATION AND BACKGROUND OF MEGHALAYA

Present Meghalaya comprising an area of 22,429 sq. km and having a population size of about 2,966,889 persons (Government of Meghalaya 2016, 3) is divided into 11 administrative districts: East Khasi Hills, West Khasi Hills, South West Khasi Hills, West Jaintia Hills, East Jaintia Hills, Ri-Bhoi, East Garo Hills, West Garo Hills, South West Garo Hills, North Garo Hills and South Garo Hills.[2] After independence, the United Khasi and Jaintia Hills district and the Garo Hills district were made parts of the state of Assam. The inclusion of these two districts under the state of Assam went against the political aspirations of the indigenous tribes that aspired for a distinct political identity of their own under the Indian Union. Their aspiration for autonomy drove them to fight for a separate hill state. After a long-drawn peaceful struggle, these two hill districts were merged to form an autonomous state on 2 April 1970, and after 21 months, it was officially declared as the full-fledged state of Meghalaya on 21 January 1972.

The state of Meghalaya covers an area of 22, 429 sq. km. It is a continuation of the Deccan Plateau, and its landscape is interspersed with hills, deep valleys, gorges and highland plateaus. The state is accessible by road from Assam; it is bounded on the north and east by the districts of Kamrup, Goalpara and Dhubri of Assam, on the south by the divisions of Sylhet and Mymensingh of Bangladesh and on the west by the division of Rangpur of Bangladesh. The state shares 443 km of the international border with Bangladesh, of which 255 km falls in various sections of the Khasi and Jaintia Hills districts and 188 km falls within the various Garo Hills districts. There are 784 border villages inhabited by a population of 210,838 in the Khasi and Jaintia Hills district areas, whereas in the Garo Hills districts there are 739 villages with a population of 198,719 (Government of Meghalaya 2010, 6–7). The border areas and villages are those areas that are within a 10-km radius from the international borders. The border areas constitute about 22 per cent of the area of the state, and their populations constitute about 23 per cent of the population of the state (Bhagawati 2007, 131). Meghalaya is predominantly a tribal state, with Scheduled Tribes constituting 86.15 per cent of the total population of the state (Government of Meghalaya 2016, 11). The state literacy rate is 74.43 per cent (Government of Meghalaya 2016, 66). About two-thirds of the state population is engaged in agriculture, and the principal crops of the state are rice, maize, potatoes, ginger, turmeric, areca nuts, etc., with various types of vegetables. Marginal and small farmers own 76.57 per cent of the agrarian landholdings (Government of Meghalaya 2016, 23). The agricultural production is low, and the state has to import food grains from outside. The state is blessed with an abundance of minerals; coal and limestone are the principal minerals, and the others such as clay (lithomargic), granite, kaolin, iron ore, quartz, gypsum, uranium, etc., are also found in the state.[3] In spite of being mineral-rich, the state did not experience industrialization. Meghalaya has only one public sector undertaking (Mawmluh Cherra Cement Limited), though in recent years a few cement production centres have come up along with some small-scale and cottage industries. Government employment and agriculture continue to be the main source of livelihood for the bulk of the people of the state.

TRADE BETWEEN THE UNITED KHASI AND JAINTIA HILLS AND SYLHET

Khasi and Jaintia Hills had traditional trade relations with the plains of what constitutes present-day Bangladesh. The main items of trade from the hills were limestone, iron, honey and wax, ivory, areca nuts, betel leaves, bay leaves, oranges, cotton, etc., whereas rice, salt, shells, gold and other precious metals were purchased or bartered from the plains. In 1790, Sylhet district had over 600 named marketplaces (Hat, Ganj and Bazaar) (Datta 2000, 208). The principal markets at the foothills on the Sylhet side were Bholaganj, Chhatak, Lakhat, Jaintia, Jafling, Pharal Bazar, Maodang, Ponatit, Sonapur, Molaghul and Lengjat. Many of these markets situated within the Sylhet boundary were frequented by Khasi and Syntheng traders. The markets or *haats* were held at a regular interval of 8 days (the Khasi week consists of 8 days), with the object of permitting the same people to visit different places on rotation.[4] Robert Lindsay, working as the resident collector in the 1770s, mentioned that the Khasi Hills produced a variety of woods suitable for building of boats and shipbuilding. He noted that the iron that was brought from the hills in lumps of adhesive sand, producing malleable virgin iron, was superior in quality compared to iron produced in Europe by charcoal. He wrote on the inexhaustible quantity of the finest oranges found growing spontaneously in the Khasi Hills, and on limestone of the finest quality, found nowhere else in Bengal or even Hindustan. He noted that the limestone trade was occupied by Armenians, Greeks and low Europeans. Lindsay himself made a fortune by way of trading in limestone.[5]

Around 1841, William Robinson, who also wrote on trade between the Khasi and Jaintia Hills and the plains, mentioned that considerable trade in clothes was carried out between the plains and the hills. He noted that Jaintiapur, the capital of the Jaintia Kingdom, was a great entrepôt in which all commercial dealings between the plains and the hills were transacted and various items from the hills were bartered for salt, tobacco, rice and goats (Robinson 1975, 408). Colonel Lister, writing in 1853, estimated that around 20,000 mounds of iron was exported from the hills to the plains of Assam to be used by boat builders for making iron clamps (Prakash 2007, 296). The close trade relations even led to matrimonial alliances between the tribes of the hills and the Bengalis in the plains (Ludden 2003). Existence of a road and river communication network between the War areas (War refers to a cultural segment of the Khasi tribe inhabiting the southern slopes of Meghalaya plateau bordering the Surma Valley) in the hills and Sylhet in the plains enabled trade, commercial and other interaction (Gassah 1988). In 1885–1886, the hills exported 85,581 tonnes of lime and limestone (Singha 1978, 392–393). Table 11.1 gives us an approximate idea of the trade in principal items.

Table 11.1 Export of Goods from the Khasi Hills in the Pre-independence Period (Tonnes)

Items	1881–1882	1885–1886	1891–1892	1901–1902
Potatoes	4,669	3,885	1,073	NA
Oranges	3,959	NA	4,218	NA
Rubber	NA	NA	NA	1.4
Cotton (ginned)	NA	NA	1,110	666
Lime and limestone	59,681	85,581	67,559	53,916

Source: shodhganga.inflibnet.ac.in/bitstream/1063/69653/19/19_chapter%2012.pdf

In the 1920s, the trade with the plains expanded further with the addition of bay leaves, betel leaves and betel nuts (Singha 1978, 393). The expansion of trade provided livelihood to many people as porters carrying goods down the slope to the plains and then carrying goods up to the hills (Gurdon 2015[1907]). The construction of a road connecting Shillong and Sylhet via Dawki in 1933 reduced the journey time between Shillong and Sylhet.[6]

INDIAN INDEPENDENCE AND TRADE RELATIONS

The transfer of power and the subsequent partition of British India into India and Pakistan led to closure and sealing of borders and border *haats* and adversely affected the lives of the hill people living in the southern slopes of Meghalaya plateau. After the trade with Sylhet stopped, the people did not know what to do with their surplus produce, especially the perishable goods such as fishes, vegetables, fruits, etc. In the absence of proper roads connecting the villages of the southern slopes of Meghalaya to the rest of the state, it becomes difficult to move goods. Some of the crumbling ruins of buildings in the border villages bear testimony to the prosperous past that characterized the southern slopes of Meghalaya plateau bordering present-day Bangladesh (Laloo 2018). The areas that were once brimming with trade-related activities are now crippled with economic stagnation, abject poverty and high unemployment. As the road and market connectivity with other parts of the state was limited, the problems of the people in the border areas multiplied after independence (Bhagawati 2007, 132).

Consequently, all cash crops—the mainstay of the border people—lost their market and buyers. The price of oranges, which ruled at ₹20–25 per *luti* (1,024 numbers), tumbled down to ₹8–16 per *luti*, and that of betel leaves, which used to sell at ₹20–30 per *kuri* (2,880 leaves), came down to ₹2–₹3 per *kuri*. *Tezpatta* and other fruits also suffered equally. At times, even at these rock bottom prices, the products did not find takers (Singha 1978, 396).

After the partition, an agreement was arrived at between India and the then East Pakistan, through a trade agreement in 1949–1950, to allow import and export of a few items, which, compared to the past, was small in both volume and type of commodities traded.[7] In the bordering areas, some items were exchanged mainly through the barter system, and there were instances of smuggling and other illegal activities. There were also frequent instances of hostility and firing on the borders, which obstructed trade activities. Finally, in December 1957, all trading activities came to a halt with stringent measures adopted by the then government of East Pakistan in the name of containing and preventing smuggling across the borders (Singha 1978, 396–397). The collapse of the economy of the border areas affected not only the traders but also the producers, agricultural labourers, mine workers and others, leading to loss of livelihoods and growth of unemployment in the hills. The prices of all such essential commodities as rice, dry fish, mustard oil, etc., increased enormously in the hills, as they could not be imported easily from the plains of Sylhet.

MEASURES TAKEN FOR IMPROVING BORDER TRADE

To deal with the post-partition crisis in the border areas, in 1952, the government of Assam appointed a high-power committee headed by H. V. R. Iyengar, the then secretary of the union home ministry (Singha 1978, 400). The committee, among other things, suggested strengthening the communication network to facilitate the transport of agricultural products of the area, subsidizing the foodstuffs, allocating funds for intensification of the agricultural programmes, etc. (Singha 1978). This was followed by the Lall Singh Committee in 1954 which suggested setting up fruit preservation plants on the border areas (Bhagawati 2007, 133). The government of Assam appointed three more committees in 1959 to suggest measures for the improvement of the economy of the border areas (Singha 1978, 400). The committees felt that the recommendation of the earlier committee could not be implemented due to insufficient funding; they suggested effectively implementing schemes, improving administration, opening more fair-price shops and making the border areas self-sufficient (Bhagawati 2007, 134).

Despite such initiatives, the ground realities in the border areas did not improve. The most common demand of the people in the border areas to open trade across the international borders could not be realized, as it was not within the competence of either the state or the union government and was dependent on the response from a foreign state that was neither friendly nor eager to open the borders for trade.

CREATION OF BANGLADESH AND BORDER TRADE ISSUES

The liberation and emergence of Bangladesh raised the expectations for the revival of trade among the people of Meghalaya, especially those residing in the border areas. The 25-year 'Treaty of Friendship, Cooperation and Peace' was concluded between India and Bangladesh on 19 March 1972. The treaty had clauses relating to security and defence cooperation and outlined the future course of economic, technical, commercial and cultural ties between the two nations (Nair 2008, 47). Subsequently, a trade agreement between the two countries was signed on 28 March 1972. It expressed the desire to promote trade and commercial relations based on mutual benefit, understanding and friendship (Nair 2008, 103).

The agreement had three important components. The first dealt with trade between two governments; it specified that Bangladesh was to export fish, raw jute, newsprint and naphtha, and India was to export to Bangladesh cement, coal, machinery and non-manufactured tobacco. The second component dealt with border trade between Bangladesh and its neighbouring Indian states of Assam, West Bengal,

Tripura, Mizoram and Meghalaya. It provided that the population residing within 16 km from the international border could enter the territory of either country once a day through specified routes for trade in specified commodities for use of border people only. The third and final component of the agreement dealt with transit facilities of goods and provided for the transport of one country's goods by waterways, railways and roadways of the other country.[8] But unfortunately, this agreement did not last even for a year; the main obstacle was that the Indian rupee could not be accepted for bilateral trade, as Bangladesh was not in favour of trade in the local currency. Moreover, free border trade led to large-scale smuggling and other criminal activities. Unable to deal with these, the two countries mutually agreed to revoke the agreement in October 1972 (Rahman 2005, 3).

The people in the border areas were disillusioned with the closure of trade, and it ended their dreams of better economic prospects. Through other agreements and negotiations, foreign trade between the two countries steadily improved the quantum of exports and imports, but that hardly made a difference. The people were left with produce that could not be lifted to the nearest markets due to the absence of a good con munication network. They had no access to warehouses and cold-storage facilities, and hence they could not store their produce for long. In the absence of skills, they could not adopt modern techniques of cropping which could enhance their income. The government of Meghalaya, to ameliorate the sufferings of the border people and to achieve development in the area, started the Border Areas Development Department in 1973, and it became a full-fledged Directorate of Border Area Development in 1975 (Government of Meghalaya 2010, 7). The department has initiated developmental activities in the border areas. It is observed that all sections of people in the state have benefitted from various schemes launched by the department.[9]

CONTEMPORARY BORDER TRADE: BALAT BORDER *HAAT* IN EAST KHASI HILLS DISTRICT

During the visit of Bangladesh's Prime Minister Ms. Sheikh Hasina to India on 10–13 January 2010, the two governments agreed to establish border *haats* on a pilot basis at selected areas, including on the Meghalaya borders, to allow trade in specified products as per the regulations agreed upon and notified by both the governments. To implement the same, a Memorandum of Understanding (MoU) and mode of operation of border *haats* across the borders between India and Bangladesh was signed on 23 October 2010.[10] The MoU stated that the objective of border *haats* was to promote the well-being of the people dwelling in remote areas across the borders of both countries through establishing the traditional system of marketing the local pro-duce through local markets. Under the MoU, both countries agreed to establish border *haats* across the border with an initial pilot project of establishing two border *haats*—one at Baliamari (Kurigram district of Bangladesh) and Kalaichar (West Garo Hills district of Meghalaya, India) and another at Lauwaghar, Dalora (Sunamganj district of Bangladesh) and Balat (East Khasi Hills district of Meghalaya, India). It was decided to identify locations for the opening of more such *haats* after taking into consideration the history, location, accessibility and interdependence of the population on either side of the border.[11] Under the MoU, both the countries decided to establish a joint committee to be headed by the joint secretary of the two countries to review, suggest modifications and propose new locations for border *haats*. The joint committee meeting was to be held annually or earlier, if decided mutually. The MoU was to remain operational for 3 years from the date of commencement of the border *haats*. The MoU could also be suspended by either party through giving an advance 30-day notice.

To facilitate smooth trade, Government of India issued a notifica-tion on 12 October 2011 permitting the following items to be traded in border *haats*: (a) locally produced vegetables, food items, fruits, spices; (b) minor local forest produce, for example, bamboo grass and broomstick, excluding timber; (c) products of local cottage industries, like *gamchas*, lungis, etc.; (d) small, locally produced agricultural house-hold implements, for example, *daos*, ploughs, axes, spades, chisels, etc.; and (e) locally produced garments, melamine products, processed food items, fruit juice, etc. The notification clarified that the term 'locally produced' implied the produce of the concerned border districts of a designated *haat*. The vendors who were allowed to sell their products should be residents of the area within a 5-km radius from the location of the border *haat* (Government of India 2011). In the subsequent months, further modalities were worked out between the administration of the two countries for the operation of the Luwaghar, Dalora (Sunamganj)–Balat (East Khasi Hills) border *haat*. Finally, the *haat* was inaugurated on 1 May 2012.

KEY FACTS ABOUT THE BALAT-SUNAMGANJ *HAAT*

The *haat* is located 98 km from Shillong, near the zero line at pillar number 1213. It has two entry points—one from Luwaghar (Bangladesh) and the other from Balat (Meghalaya). The *haat* is managed by a management committee constituted by the two countries for the management of their respective areas. The management committee comprises five members headed by the Additional District Magistrate/ Sub-Divisional Magistrate (SDM) of the district having jurisdiction over the designated *haat*. Either country has one representative each from the police, customs agency, border security agency and village/union local government. The Additional District Magistrate/SDM may authorize an officer from the local administration for day-to-day monitoring of the *haats* and handling of emergencies. The committee maintains the list of authorized vendors/vendees, suggests items for trade and enforces health precautions. It also holds joint meetings to discuss operational issues as and when required, after giving an advance 2-day notice. The border *haat* is open once in a week, on every Tuesday, from 10 a.m. to 3 p.m. The timing and frequency can be modified/set by the mutual consent of the border management committees.

The number of vendors is limited to 25 from each country. The list of vendors is exchanged by the respective *haat* management committees. The vendors are selected by the committee initially for 1 year and comprise those who are residents of the areas that fall within 5 km from international borders. They are provided with photo identity cards and can enter the market from the respective entry point of their country. Only residents of the area within a 5-km radius from the location of the border *haat* are to be allowed to buy products from the *haat*. The number of vendees is regulated by the respective *haat* management committees, so that the *haat* would not be overcrowded. The vendees are provided with identity cards.

The *haat* management committee and border security agencies maintain entry/exit records of the vendors and vendees. At the time of closing of the *haats*, the records are checked to ensure that every person who entered the *haats* has returned to their respective country. Security forces personnel are not to be ordinarily allowed inside the *haats* except in case of emergency; however, there is no restriction on the presence of security personnel outside the boundary of the *haats*. Both the Indian rupee and the Bangladeshi taka can be used for purchase inside the border *haats*; however, the maximum value of an individual purchase cannot exceed US$50.00 (now revised to US$100.00).

FINDINGS FROM THE FIELD SURVEY

The following facts emerged from our observation in the market, focus group discussion and interaction and discussion with vendors, vendees, custom officials, security personnel and K. N. Syiem, the Syiem of Bhowal, during our visit to the market in January and February 2017.

VENDORS, VENDEES AND COMMODITIES

Although only 25 vendors are to be allowed from either country, the management committee of the *haat* selected 50 vendors from the Indian side. The decision was taken because of the large number of applications. It was decided that if the vendors with serial numbers 1–25 were allowed in the first week, the vendors with serial numbers 26–50 would go next week, and the rotation would continue. In the selection of vendors, representation was given to all major ethnic populations residing within the border area, namely Khasis, Bengalis, Garos, Hajongs, etc. All the vendors were from surrounding areas, such as Dangar, Dangardombah, Mawpen, Kynrang, Laitumsaw, Dewsawlia, Ingkyrsa, Pyndensohsaw, Kynrang, Mawpen, Khagorkora, Pyndenborsora, Dangardop and Balat. We were also informed that the vendors of Bangladesh were also from the adjoining areas. The visiting team noted that no female vendor was present from Bangladesh. It was also observed that in addition to the 25 vendors occupying the Bangladesh-demarcated portion of the *haat*, quite a few vegetable vendors from Bangladesh were selling their vegetables in makeshift arrangements. No such additional vendor was present on the Indian side of the market.

It was noticed that several buyers from adjoining villages were entering the market. The entry/exit points on the Indian side are maintained by the Border Security Force (BSF). The identity cards of the people entering the *haat* are checked, and details of the visitors are entered in the register. It was observed that representatives of customs and local village leaders, including K. N. Syiem, the Syiem of Bhowal, were assisting the BSF officials in managing the crowd entering the *haat*. At the Bangladesh entry point to the market, Border Guards Bangladesh (as the border police are known in Bangladesh) were checking the passes of the vendees. Only 150 vendees are allowed to enter from either side at any given time. Those who enter take a long time to exit, as there is no maximum duration imposed on the vendees; therefore, many of them cannot enter the market despite waiting for long. We were also informed that every week, several visitors from Shillong and elsewhere in substantial numbers visit the *haat*. As a result, many of the genuine buyers from border villages are deprived of the opportunity of entering the market. Many of the vendees we found make big purchases for business purposes and not for personal use.

Apart from the items of local produce as specified in the list mutually accepted by both countries, the team found that many of the items that cannot be called local products of the border districts/areas were also sold in the *haat*. The Bangladeshi vendors were found selling such items as garments, melamine, drinks, snacks, barring bakery items, and food grains brought from Dhaka, Chittagong, Narayanpur, etc. It was also found that some of the items sold were produced in a third country. The Indian vendors were selling such items as blankets, health drinks, chocolates, snacks, spices, dry fruits, etc. While some of these items were from the bordering district, there were also items from outside the state or from multinational companies. The only products that we found local in the *haat* were bay leaves and oranges.

CURRENCY AND VOLUME OF TRADE

Both the Indian rupee and the Bangladeshi taka are accepted for trade; trade can also be done through barter. The Indian vendors can exchange the Bangladeshi currency in the UCO Bank branch in Dangar, Balat, and we were informed that a similar arrangement exists with Sonali Bank for the Bangladeshi vendors. It was also found that sufficient currency, in both Indian rupees and Bangladeshi takas, is held by the vendors, and the buyers do not face any problems in purchases. The limitation of US$100.00 seems sufficient for individual buyers; many villagers informed that they purchase ₹1,000–1,500 worth of goods when they visit the *haat*. It was difficult to ascertain precisely the volume of trade, as the vendors on both sides were hesitant to give the figures. Trade seems to be brisk, as was evident from the head loads being carried from the market. Most of the big purchases from Indian buyers were snacks, fruit juice, melamine dishes and different plastic items for domestic use. Most of the items that Indian vendors were selling included blankets, spices, biscuits, health drinks, etc. The vendors admitted that normally each vendor could make an average profit of ₹5,000 on every *haat* day.

SECURITY AND OTHER RELATED ISSUES

The entry/exit registers are very rigorously maintained by the BSF. The crowd management on the Indian side appeared better compared to that around the entry/exit points on the Bangladeshi side. One sees a bit of chaos and use of force by the Border Guards Bangladesh to regulate the crowd. No mobile phone is allowed inside the market, and photography is prohibited.

The opening of the border *haat* has generated ancillary employment opportunities and provided livelihood to the people of Balat and surrounding villages. There are quite a few porters who carry the goods that are to be sold in the market, as well as carrying the big purchases from the market, as no vehicle is allowed beyond the border fencing. The goods need to be carried either through handcarts or as head loads; this has generated employment for 20-odd people, and they earn an income of ₹400–500 on a *haat* day. Moreover, some snacks and tea stalls have been set up at the entry point outside the fencing on the Indian side, which provide additional income to a few families during the *haat* days. The *haat* has also generated revenue for transporters, as a huge number of commercial vehicles ply for carrying goods and passengers on *haat* days.

SUGGESTIONS FOR IMPROVEMENT

Based on the study, one can make the following suggestions for improving the border trade:

Connectivity: Balat, located 98 km from Shillong, is not physically connected well; the roads in most places are in a deplorable state, with broken bridges and rough patches. There is no metal road from Dangar to the *haat*, for a distance of about 4 km. There is a metal-surfaced road that runs parallel to the border fencing, but it is not allowed for civilian vehicles. During the rainy season, it becomes impossible for a vehicle to reach the entry point of the *haat*. If the connectivity is improved, there would be an increase in trade.

Infrastructure: The infrastructure of the *haat* needs improvement. There was no functioning toilet in the market, nor was there any provision for drinking water. Initially, it seems, there was one gents' and one ladies' toilet, but they are not functioning at present. The market space can also be broadened, as the present 75 m × 75 m area with 50 vendors and 300 vendees makes the *haat* extremely crowded. There is not much scope within the given space to add other infrastructural facilities. It would be beneficial, especially for children and the elderly, if some shed is built for resting. It could also serve to promote contact among people and lead to better understanding. There could also be a first-aid room in the *haat* to deal with medical emergencies.

Commodities for trade: Barring agricultural products, most products traded in the *haat* are not products of the adjoining districts. They are brought from other parts of the country and in some cases even from a third country. Therefore, for such products, the government should levy some small duty to discourage any big trade in such items. Barring such items from trade would not solve the problem but lead to illegal trade as long as the demand remains. Limits can be fixed on the number of items purchased from the market, so that the real purpose of the *haat* is served. It seems there is a high demand for fish and poultry items among the villagers in the border areas of India. These items are expensive, as they are mostly procured from Shillong, involving a high cost of transportation, and there is also some illegal trade in these items across the borders. If these items are allowed to be sold in the *haat*, the villagers could have the advantage of cheaper rates. Therefore, it is necessary to build quarantine and other facilities so that these items can be traded in the *haat*.

Entry and exit: It was seen during our field survey that many people were not able to enter the market, as at no point could more than 300 buyers enter, and many of those who entered in the morning remained there till the end of the *haat's* functioning hours, taking advantage of the absence of time restrictions. A similar crowd was also seen outside the entry point from the Bangladeshi side. Therefore, there is a need to increase the number of persons to be allowed in the market, as well as to impose some maximum time for which an individual buyer can be allowed to remain inside the *haat*, so that more genuine buyers can enter the *haat*.

Increase in the number of vendors/timing of *haats*: It seems that there is a huge demand for vendors in these *haats*. In this light, 25 from each country seems a small number; having more vendors could also lead to better and more competitive trade. Further, traders may be permitted to trade in one or two specific items instead of dealing with many items. In terms of the distribution of vendors, during our visit, we noticed that there were only a few female vendors on the Indian side, and there were none on the Bangladeshi side. The number of vendors from either side can be mutually increased to 50. Similarly, the timing of the *haats* can be increased, as life starts early in this part of the country. It could start from 8 a.m. and during the summer months be extended till 4 p.m.

SUMMARY

Meghalaya, like India's other north-eastern states, has been lagging behind the rest of India in economic growth and development. Because of its geographical remoteness, lack of basic infrastructure and negative publicity on militant activities and social strife, the state could not emerge as an attractive destination for foreign direct investments and even investments by domestic industrial houses. There is hardly any effective programme aiming at improving production or increasing the income-generation capacity of the people in the border areas. The central government has indeed been generously funding some schemes, but then it has only led to further expansion of the already inflated government sector. The majority of the people of the state are engaged either in the primary sector or in the service sector. In this context, under the present circumstances, border trade carries hope for almost a quarter of the population of the state living in the bordering villages to get out of the vicious cycle of poverty. A nation must indeed have well-demarcated borders, but there is also a need to consider the interests of the people living across the borders. By and large, the experiment of trading at Balat has been successful. The villagers in Balat and adjoining areas are now able to purchase items not locally available at much cheaper rates and avoid travelling to Shillong, across a distance of 98 km, and paying a higher price for them. As K. N. Syiem correctly pointed out,

> Since the opening of this haat on 1 May 2012, the people within the 5 km radius from this border haat and even beyond that have very much benefited.... We are glad that the Government of India and Bangladesh have agreed to open four more border haats in Meghalaya.... (Syiem 2016)

NOTES

1. See http://www.carecinstitute.org/uploads/docs/Cross-BorderTrade-CAREC.pdf (accessed 21 February 2017).
2. See http://www.meghalaya.gov.in/megportal/stateprofile (accessed 22 February 2017).
3. See http//:www.megadmg.gov.in/minerals.html (accessed 21 February 2017).
4. See http://www.yellowbaobab.blogspot.in/2013/02/shillong-anglican-church.html (accessed 1 March 2017).
5. See http://www.yellowbhaobab.blogspot.in/2013/01/lives-of-lindsays.html (accessed 1 March 2017).
6. Shullai quoted it from the address of Sir Michael Kean dated 22 March 1933. See Shullai (1980).

7. See https://www.commonlii.org>other>treaties>INTSER and www.commonlii.org>Databases (accessed 9 January 2020).
8. See http://www.mea.gov.in/bilateral-documents.htm?dtl/5606/TradeAgreement+Protocol+1+Nov+1972 (accessed 21 February 2017).
9. See https://www.niti.gov.in/niti/writeraddata/files/document_publication/report-BADP.pdf (accessed 9 January 2019).
10. See http://www.pib.nic.in/newsitc/Print/Release.aspx?relid=134202 (accessed 27 February 2017).
11. See http://www.commerce.nic.in/trade/MOU-Border-Haats-across-Border-India-andBangladesh2010 (accessed 17 February 2017).

REFERENCES

Bhagawati, Dhiren. 2007. *Meghalaya, Issues, and Legacies of Its Early Years*. Guwahati: DVS Publishers.

Datta, Rajat. 2000. *Economy and Market: Commercialization in Rural Bengal*. New Delhi: Manohar Publishers and Distributors.

Gassah, L S. 1988. *Trade Routes and Trade Relations Between Jaintia Hills and Sylhet District in the Pre-Independence Period*. NEIHA Proceedings, 9th Session, Guwahati.

Government of India. 2011. Instruction No. 2 of 2011 dated 12th October. Shillong: Ministry of Finance, Department of Revenue, Office of the Commissioner of Customs (Preventive) North Eastern Region.

Government of Meghalaya. 2010. *Report of the Directorate of Border Areas Development*. Shillong: Border Area Development Department.

Government of Meghalaya. 2016. *Statistical Abstract Meghalaya 2016*. Shillong: Directorate of Economics and Statistics.

Gurdon, P. R. T. 2015[1907]. *The Khasis*. London: Palala Press.

Laloo, Sashi Teibor. 2018. 'My Meikha/ Grandmother's Tales of the Partition from Khasi-Jaintia Hills'. *Raiot*, 17 May. http://www.raiot.in/my-meikha-grandmothers-tales-of-the-partition-from-the-khasi-jaintia-hills/ (accessed 8 February 2021).

Ludden, David. 2003. 'Investing in Nature around Sylhet: An Excursion into Geographical History'. *Economic and Political Weekly* 38(48):5080–5088.

Nair, Sukumaran P. 2008. *Indo-Bangladesh Relations*. New Delhi: APH Publishing Corporation.

Prakash, Ved. 2007. *Encyclopedia of North East India*, Volume I. New Delhi: Atlantic Publishers and Distributors (P) Ltd.

Rahman, Mohammad Matizur. 2005. 'Bangladesh–India Trade: Causes for Imbalance and Measures for Improvement.' http://www.eprints.usq.edu.au/4196/1/Rahman_JABE_v8nl.pdf (accessed 21 February 2017).

Robinson, William. 1975. *A Descriptive Account of Assam*. New Delhi: SanskaranPrakashak.

Shullai, L. G. 1980. 'Umtyngar-Jaintiapur Road'. *Ropeca*, Shillong, 27 August. Cited in Gassah, L.S. 'The War Jaintias', Department of Arts and Culture, Government of Meghalaya, http://megartsculture.gov.in/herit-war-jaintias.htm (accessed 21 April 2021).

Singha, Jagadish Chandra. 1978. 'Socio-Economic Development of the United Khasi Jaintia Hills and Garo Hills Since Independence.' Unpublished PhD Thesis. Guwahati: Department of Economics, Gauhati University.

Syiem, K. N. 2016. Speech at the Meeting of Consultation Meet on 'Border Haats and their Socio-Economic Impact', Guwahati, 26 October.

PART II

Border Crossing and Inter-Community Relations: Cooperation and Conflict

Chapter 12

Nepali/Gorkha Settlers in Northeast India
Colonial Encounters and Post-colonial Dilemmas

Tejimala Gurung Nag

INTRODUCTION

Several lakhs of Nepalis[1] or Gorkhas[2] are settled in different north-eastern states. According to the 2011 census, Assam has over 5 lakh Nepali-speaking people, with Sikkim, Manipur and Meghalaya having Gorkha populations of 382,200, 63,756 and 54,716, respectively. Similarly, we find Gorkha settlements in Arunachal Pradesh and Nagaland. Nepali-speaking people account for 63 per cent of the population in Sikkim. The Gorkhas are mostly soldiers, agriculturists and herders, while some have emerged as professionals and businesspeople. They have their own cultural and political associations, and in Assam there are/were Gorkha Members of Legislative Assembly (MLAs) and Members of Parliament (MPs). Though they are Indian citizens, there is a misconception among the local population in the north-eastern states that all Gorkhas are illegal immigrants from Nepal who have entered India by taking advantage of the Indo–Nepal Treaty of Peace and Friendship 1950. Such misconceptions have often led to tensions between the dominant local communities and the Gorkha population (Gurung 2002, 149–158). Against the background of these facts, an effort is made in this chapter to present an overview of the rise of Shah rule in Nepal and Nepal's encounters with the British, and explain the migration and settlement of the Gorkhas in the north-eastern states in particular.

EMERGENCE OF THE SHAH DYNASTY IN NEPAL

Nepal[3] is a country of great ethnic and linguistic variety, with 70 languages or dialects across a population of approximately 26 million at present.[4] Besides Hinduism and Buddhism, tribal religious and shamanic traditions are practised in Nepal. Nepal has three major areas: the High Himalayas (bordering Tibet and China), the Middle Hills, which form the cultural and political heart of the country, and the Terai plains[5] on the south bordering India. The three major river systems of Nepal flowing from west to east are the Karnali, the Gandaki and the Kosi. The physical terrain that produced variations in climate and soil has also preserved ethnic and cultural differences. Hemmed in by the High Himalayas and China in the north and by India on three sides, Nepal's political geography and environment, to a considerable extent, conditions its relations with India. Till the middle of the 18th century, Nepal was divided into several small principalities. Gorkha was one such principality ruled by the Shah dynasty. Prithvi Narayan Shah, who ascended the throne of Gorkha in 1743, united all principalities, including the Kathmandu Valley, either through war or diplomacy, and founded a powerful Gorkha kingdom in 1767, invoking Hindu religion. He shifted his capital to Kathmandu and ruled the kingdom that came to be known as Nepal. Prithvi Narayan Shah realized that his kingdom was a 'yam between two rocks' (China and India) and hence had to tread a realistic path. He strengthened his defence forces and chose not to antagonize the powerful neighbours—China and India. After his death in 1775, his sons and successors continued to extend the boundaries of Nepal.

ANGLO-NEPAL WAR, DELINEATION OF BORDERS AND CREATION OF DARJEELING

At about the same time when the Shah rulers were expanding the boundaries of their kingdom in Nepal, the English East India Company was consolidating its rule in India. The emergence of Nepal as a united entity under the Shah dynasty and the coterminous establishment of British rule in India during the 18th century soon brought the two expanding powers into conflict. For the British, the Himalayan region comprising Nepal, Sikkim and Bhutan became strategically important from its perception of imperial security and defence. The British interest in the Himalayan region also resulted from an increasing concern with Russian and Chinese activities on the Himalayan border during the colonial period. The British, who were looking for expanding trade to Tibet and China, viewed Kathmandu Valley as a natural entrepôt for trade between India and Tibet. Taking advantage of the British weakness and political fluidity in North India, the Gorkha rulers subordinated the neighbouring hill kingdoms of Kumaon (1790), Garhwal (1803) and Sirmour (1805). However, Nepal's aggressive policies towards Tibet and Sikkim, and its southward expansion threatening the British interests, led to the Anglo-Nepal War in 1814. The British, which by then had consolidated its rule in India, halted Nepal's aggressive policy of expansion and compelled the Gorkha ruler to sign the Treaty of Sugauli in 1816, forcing Nepal to cede Garhwal, Kumaon and the Terai to the British East India Company. The Anglo-Nepal War (1814–1816) provided the British with the opportunity to demarcate the Indo-Nepal border and put an end to the problems arising out of the fluid, overlapping systems of control which characterized the Terai or the lowlands through establishing an unambiguous frontier. The Indo-Nepal border was, however, kept open by the British for trade, to procure valuable forest resources and to facilitate recruitment of Gorkha soldiers.

Further towards the east, Nepal had to relinquish territories occupied from Sikkim, which included the entire lowlands between the Mechi River and the Teesta River and all the territories within the hills situated eastward of the River Mechi. By the Treaty of Titalia entered between the East India Company and Sikkim (10 February 1817), these were handed over to Sikkim. The British search for a sanatorium in the hills subsequently led to the growth and development of Darjeeling as a colonial hill station. In 1839, the Raja of Sikkim under a 'deed of grant' handed over Darjeeling, then a small settlement of about 100 people, to the British. This was subsequently developed into a larger district through further incorporation of tracts from Sikkim and Bhutan. The British encouraged more numbers of

Nepali settlers from the contiguous Limbuwan area to Sikkim and Darjeeling for development of the nascent tea industry and for imperial strategy. The Treaty of Tumlong (March 1861) made Sikkim a subservient state.[6] Darjeeling, which had formed part of Sikkim, Bhutan and Nepal at various points of time before being taken by the British, contained a heterogonous population of Lepcha, Bhutia, Limbu, Rai, Magar, Tamang and Koch people. When Darjeeling was added to British India, it carried this mixed demographic structure. At present, the Indo–Nepal border[7] extends for a length of 1,751 km, from Uttarakhand in the west to Darjeeling (the River Mechi) in the east, sharing borders with five Indian states (Government of India 2008). With Sikkim and Darjeeling, Nepal shares a border of about 197 km (Lama and Khawas 2009). The Indo–Nepal border, which is an open border, is a colonial legacy.

BRITISH–NEPAL RELATIONS AFTER THE SUGAULI TREATY

Following the Treaty of Sugauli, restrictions were imposed on Nepal's external dealings with other powers, and Nepal had to agree to have a British resident at Kathmandu. Despite anti–British sentiments among the Nepalis, Bhim Sen, the prime minister (1816–1839), cooperated with the British to save Nepal from losing her independent status at the hands of the East India Company. From the 1840s, effective political power in Nepal shifted from the Shah kings into the hands of the Rana family that served as hereditary prime minister from 1846 to 1951. In 1856, Jang Bahadur Rana secured, by a royal decree from the king, de facto sovereign power with the official title of maharaja for the prime minister and broad supervisory powers over the king of Nepal, who henceforth came to be called the Maharajadhiraj. The maharaja prime minister became the virtual ruler, with the king as a figurehead. Nepal under the Ranas was a military state with power concentrated in a family that wanted to keep Nepal in a feudal state aloof from modernizing democratic elements. In 1848, Jang Bahadur Rana had offered six regiments of Nepali troops to the British as reinforcement for the Second Anglo-Sikh War. Nepal provided troops and logistical support to the British to suppress the 'Sepoy Mutiny' in 1857. In return for the favours, the British, through the Treaty of 1860, restored to Nepal the whole of the lowlands or the Terai extending eastwards from the Kali River to the foot of the hills north of Bagowra Tal in the district of Gorakhpur. For his help, Jang Bahadur Rana was also awarded an honorary knighthood by the British. Jang Bahadur Rana was eager to establish good relations with the British to restrain the activities of his rivals living in exile in India. However, in internal matters, he continued the 'closed door' policy, refused to allow European traders, rejected a proposal to construct a road from India to Kathmandu and allowed the British staff and the resident to visit only limited areas (Whelpton 2010, 47). For the British, a dependent and obliging family oligarchy suited their political interests of maintaining peace and security in their frontier. The British did not interfere in the internal affairs of Nepal but ensured that the Shahs and later the Ranas rendered support to the British in its imperialist ventures in Tibet.

COLONIAL CONNECTIVITY PROJECTS, TRADE AND INDUSTRIAL DEVELOPMENT

The construction of Indian railways up to Nepalganj, Biratnagar, Janakpur and Birganj, all bordering Nepal, enabled Nepal to import items from India, like kerosene for lamp and cooking. Indian salt replaced slowly the Tibetan salt that had for centuries been obtained through salt–grain exchange networks between Tibet and Nepal (Whelpton 2010, 76). Brian Houghton Hodgson, who served as resident (1824–1843), had emphasized developing trans-Himalayan trade to China through Tibet—in particular through an overland route from Calcutta to Peking (Waterman 2004, 7). After the British established control over Sikkim in 1860, it developed the route to Tibet through Chumbi Valley. Kalimpong had been well connected through trans-Himalayan trading networks for centuries, and the historic Kalimpong–Lhasa trade route was the shortest route from India to Tibet. With the completion of the Calcutta–Darjeeling railway in 1881, the route to Lhasa from Calcutta took only 3 weeks, reducing the journey by half through Kathmandu. These alternative connectivity projects had an adverse effect on Kathmandu, considered till then as an entrepôt of trade across the Himalayas. The final death blow came in 1904 when the British forced Tibet to permit unrestricted trade.

In the 1930s, to counteract growing public resentment against the Ranas, Bhim Shamsher and Juddha Shumsher Singh (1931–1945) sought to undertake some economic development of the country. However, nothing substantial could be achieved due to the feudal character of the Rana rule. Further, industrial development in Nepal was dependent on the support of the British, as raw materials had to be brought across Indian territory. Because of the Indian national movement, the British administration was primarily focusing on protecting the industries in India, and Nepal did not receive the expected support from the British. It was evident that as Nepal's economy became more and more bound up with India, it would become economically dependent on India.

RECRUITMENT OF GORKHAS IN THE BRITISH ARMY

During the Anglo-Nepal War (1814–1816), the British used Gorkha defectors, comprising mainly Kumaoni, Garhwali and Sirmouri hill men as irregular levy. Impressed by the bravery of the Gorkha soldiers in the war, the British recognized them as a martial race and started recruiting the Gorkha defectors irregular forces in the British Indian Army. In 1815, the British raised two Gorkha battalions, then called the Nussuri Battalion. Later, it was renamed as the Gorkha Rifles. The recruitment of Nepali soldiers into the Indian Army was facilitated by the system of annual rotation of troops for military service. The practice of the off-the-roll rotation system provided an opportunity for many Nepalis to gain experience in the military. Whenever the need arose, the Sikh kingdom of Punjab and the British Indian Army recruited this trained reservoir of Gorkha soldiers. Brian H. Hodgson, the British resident at Kathmandu in Nepal, was a forceful advocate for recruiting Gorkhas. According to Hodgson, in 1832 in Nepal, there were about 30,000 *dhakres* (term for Nepali soldiers off the roll on rotations) belonging to the Khas (high-caste Chettri and Thakur), Magar and Gurung tribes (Bhanskota 1984, 48). It was argued that apart from weakening Nepal militarily, the recruitment of Gorkhas in the British Army would enhance British military strength, neutralize the influence of the homogenous high-caste Bengal Army of the East India Company and provide a safety valve in times of emergency. By the 1870s, there were five Gorkha regiments in India, each with a single battalion with regimental headquarters at Dharamshala, Dehradun and Abbottabad. This happened despite the Nepal government's steadfast refusal to allow British recruiters to set foot in the country. Some Gorkha soldiers who returned to Nepal were put to death, and their property was confiscated (Mojumdar 1975, 42–43). The situation, however, changed after 1885 when the Ranas emerged as powerful prime ministers, relegating the Shah kings to nominal heads.

After 1885, Shamsher Singh Rana, the prime minister of Nepal, allowed and regularized the recruitment of Nepali soldiers termed 'Gurkhas' by the British in the British Indian army.[8] Till then, recruitment of Nepali soldiers to the British Indian Army had been carried on fraudulently by the British. After 1885, recruitment depots were opened at Gorakhpur and Darjeeling near the Nepal border, and regiments were allowed to send recruiting parties (which did not include British officers) into the interiors. For the British, the most valuable resource that Nepal could provide was the Gorkha soldiers who would fight its imperial wars and maintain its empire. It was essential for the British to ensure that the recruitment of the Gorkhas continue to be unhindered. British dependence on the Rana government had increased during the period of First World War (1914–1918) for the supply of Gorkha soldiers for fighting Britain's imperial war overseas and also for internal security within India owing to the threat posed by intensification of the nationalist movement during the period. Out of over 1 lakh Gorkhas mobilized during World War I, one-tenth were killed or wounded or missing in action (Bhanskota 1984, 126). In the aftermath of World War I, a treaty of friendship was signed on 21 December 1923 between British India and the Rana government of Nepal. The treaty was important for Nepal, as the British recognized the internal and external independence of Nepal. The new treaty contained provisions of trade and transit which allowed Nepal to import from or through British India—the arms, ammunition, machinery, warlike materials or stores it required, and it also waived customs duty at ports in British India on the goods imported by the government of Nepal.

Under the rule of Juddha Shumsher Singh, Nepal provided in all 150,000 recruits for the Indian Army and over 50,000 for the Indian military police. In 1935, it was reported by the British Envoy at Kathmandu that ₹50 lakh was paid annually as pensions to the Gorkhas and about ₹125 lakh as salaries to the Gorkha soldiers in India. During 1939–1945, over 2 lakh Gorkha soldiers were serving the British forces (Mojumdar 1975, 13). Nepal was considerably benefitted by the remittances sent and savings brought by the Gorkha soldiers employed in the British Army. When Burma fell to Japan in 1942, the British even organized the rehabilitation of thousands of Nepali refugees in rehabilitation centres in India to prevent the entry to Nepal of those Gorkhas who were of 'mixed parentage' and who had become 'lax' about Nepali customs and religion, who could upset the political and social situation in Nepal (Gurung 2016a, 203–220). This was done keeping in view the concerns expressed by the maharaja of Nepal. The Rana rulers did not want the Gorkha soldiers to return to Nepal, as they were afraid that the returnees exposed to a new environment and culture would turn against the feudal rule in Nepal. The British acknowledged the role played by the Gorkhas as part of the allied forces. The Gorkha battalions, which comprised about a fifth of the total number of Indian infantry units during the Second World War, won no less than 10 of the 26 Victoria crosses awarded to the other ranks of the Indian Army.

THE GORKHA REGIMENT AND THE NORTHEAST INDIA CONNECT

The colonial army not only served as an instrument of conquest and apparatus of rule for the British but also assisted its civil administration in the maintenance of internal security. Through the use of the armed forces, the British were able to erode the military and political authority of the native ruling states, subjugate the frontier tribes and thereby establish a monopoly of power throughout India (Alavi 1995; Cohen 2004; Omissi 1994). The military was central to the development of the British political expansion and sovereignty in India.

Apart from using them in colonial conquests and wars, the British used the Gorkhas for maintaining internal security in the country. One such colonial military instrument in the territorial expansion and consolidation of British rule in the hill areas of Northeast India was the 8th Gorkha Rifles (Gurung 2016c, 380–389). The 8th Gorkha

Rifles (comprising two battalions) was raised in the early 19th century primarily for service in Assam and to safeguard the then eastern frontier of Bengal during the outbreak of the First Anglo-Burmese War (1824–1826) (Huxford 1952, 5–10). It was engaged in various 'small wars'[9] for frontier campaigns and for many punitive expeditions against the hill tribes such as the Khasis, Jaintias, Garos, Nagas, Lushais and Abors. The British officers felt that compared to the regiments comprising high-caste soldiers from the United Provinces, the regiments like the 43rd and 44th Sylhet Light Infantry (or the 8th Gorkha Rifles), which had Nepali recruits, were more useful, as they had better experience in jungle warfare. To secure more Gorkha recruits for its pursuits in the North Eastern Region, the British even prevented the 'martial Gorkhas' from getting employed and recruited in industrial services, like in the coal mines of Assam (Gurung 2009, 265–270).

The colonial construction of the martial-race theory during the 19th century also subsequently led to the recruitment of numerous Gorkhas from eastern Nepal in the Assam Military Police in Assam (popularly known later as the Assam Rifles) and thereafter in the Burma Military Police in Burma (Gurung 2015, 520–529). The 'Cachar Levy' raised by Mr Grange, the civil charge of the Nowgong district, in 1835 formed the earliest embodied unit of the Assam Rifles (Shakespear 1977, 9). The purpose of a 'levy' or militia was to be 'a cheap semi-military body' to carry out arduous duties of guarding the eastern frontier of Assam, often involving fighting in jungle localities. The Assam Rifles worked in close cooperation with the Gorkha infantry group of the Indian Army. It was affiliated to the Gorkha Rifles of the Indian Army in 1924. By 1920, there were five battalions of the frontier military police in the Assam Valley and Hills. These were the Naga Hills Military Police Battalion (3rd Assam Rifles raised in 1835), Lushai Hills Battalion (1st Assam Rifles raised in 1863), Lakhimpur Military Police Battalion (2nd Assam Rifles raised in 1864), Darrang Military Police Battalion (4th Assam Rifles raised in 1913) and the 5th Assam Rifles raised in 1920 for duty on the Darrang and Kamrup borders. In a battalion strength of 800, the Gorkhas made up half the number. During the First World War, the Gorkhas belonging to the military police also responded to a call for volunteers to join the army, and about 14 non-commissioned officers and 500 sepoys had been drafted into the Gorkha regiments at the front.[10]

After retirement, many of these Gorkha soldiers, instead of returning to Nepal, settled down in the North Eastern Region with their families. Small pockets of Gorkha settlements thus emerged near the regimental headquarters and military police outposts. The British encouraged the Gorkha soldiers to colonize and settle down in Assam, then a densely forested and sparsely inhabited region. The Gorkha soldiers during the colonial period were thus the pioneer migrants and settlers in the region. Once a network of Nepali settlements had been established in the region, migration flowed towards the areas where the kin and associates of the migrants were located, in chain migration, resulting in migration fields or clustering of the migrants in various locations and areas in the plains and hills of Northeast India (Gurung 2017, 229).

MIGRATION OF OTHER CATEGORIES OF NEPALI MIGRANTS TO NORTHEAST INDIA

The settlement of retired Gorkha soldiers in the hills and plains of Northeast India paved the way for later migration of other categories of Nepali migrants. From the last quarter of the 19th century, herders and poor agricultural farmers also began to move across from the hills of eastern Nepal to the contiguous regions of the Himalayan foothills towards southern Bhutan and Assam. Migration of the grazers was encouraged by the prospect of dairy farming as a viable form of livelihood in the Assamese plains, in the interiors of the Khasi, Garo and Mikir hills, etc. (Gurung 2008, 159–169). Realizing the economic implications of the grazer migration and settlement, in 1915, the British appointed J. W. Arbuthnot, Special Officer, for the purpose of reporting on grazing activities in Assam. Grazing reserves were specifically demarcated for herders by the British. By 1930, this category of migrants gradually spread to almost all the districts of Assam and in the erstwhile Balipara and Sadiya Frontier Tracts (Arunachal Pradesh). The grazing reserves of Darrang district were the major attraction for Nepali herders.[11] The herders mainly belonged to the so-called non-fighting castes, and most came to stay semi-permanently, 'bringing their wives and families'. All the migrants in the Sadiya Frontier Tract came from eastern Nepal. They comprised the Brahmans and Chettris of the 'non-martial stock', the Lamas, Rais and Limbus and a fair proportion of Magars and Gurungs from eastern Nepal, and also the pensioners from the 2nd Battalion, Assam Rifles.[12] By 1901, the Nepali-speaking population in Assam increased to 21,347, which constituted 0.35 per cent to the total population. Unlike in the soldier recruitment and migration, which were sponsored by the colonial state, the economic factor lay behind the migration of grazers and marginal farmers, which was thus voluntary and not sponsored by the British.

NEPAL ON THE EVE OF THE INDIAN INDEPENDENCE

Nepal under the Ranas was a military state with power concentrated in a family that wanted to keep Nepal in a feudal state aloof from modernizing democratic elements. The British, for its own strategic and economic interests, patronized the Rana rule. However, on the eve of the Indian independence, Nepal mattered little to the British. The only treaty agreement entered was the Tripartite Agreement signed on 9 November 1947 between Britain, India and Nepal regarding the continuation in service of the Gorkhas in either the British or the Indian Army.[13] The hurried transfer of power to India by the British left Nepal with little time to address the future relations with independent India. The Ranas, facing increasing discontentment at home, wanted to clinch a deal with India to preserve Nepal's independence and security of their rule.

In 1950, with the occupation of Tibet by China, the nationalist leaders in India realized the strategic significance of the colonial policy of keeping Nepal a buffer between India and China, as evident from Jawaharlal Nehru's statement in the Indian Parliament in December 1950 (Muni 1973, 13–14). The Rana rulers were also scared of a Chinese-sponsored communist revolution that could overthrow their autocratic rule in Nepal. The identity of interests led the two countries to the conclusion of the Treaty of Peace and Friendship (1950). The Treaty of 1950 was in a way a reiteration of the Treaty of 1923. Apart from acknowledging the sovereignty of Nepal and India, the treaty assured import of arms, ammunition and security equipment of Nepal from or through India (Articles 1 and 5) and assured economic concessions and free movement of its nationals (Articles 6 and 7). Subsequently, in 1951, with the help of the Indian government, the Nepali Congress put an end to the Rana rule and restored King Tribhuvan as ruler of Nepal. While India is keen on ensuring that Nepal is under its hegemonic influence, on its part, in recent years, Nepal has been pursuing the policy of equal distance from India and China.

CONCLUDING REMARKS

There was already a considerable Nepali/Gorkha presence and settlement in the Northeast at the time of the Indian independence.[14] The Nepalis/Gorkhas were either serving in the Gorkha regiments and military police battalions or were earning a living as herdsmen or cultivators or through small-time trade and artisan services. The Tripartite Agreement signed on 9 November 1947 between Britain, India and Nepal allowed the continuation in service of the Gorkhas in either the British or the Indian Army. After the Indian independence, the Indo-Nepal Treaty of Peace and Friendship 1950 recognized what was a de facto situation through providing for an unrestricted reciprocal flow of nationals and goods of both countries, with privileges to reside, own property and take part in economic activities. Migration of Nepalis to India's Northeast, therefore, continued even in the post-independence period. Recruitment of Gorkhas in the Indian Army and also in the Assam Rifles continued alongside the migration of herdsmen. In 1951, there were 122,823 Nepali speakers in the Northeast, constituting 1.39 per cent of the total population. Nepali herdsmen migrated in great numbers to the hill districts of Assam, particularly into the Khasi and Jaintia Hills. In the United Khasi and Jaintia Hills district, the Nepalis numbered 19,721, or 5.42 per cent of the total population of the district. The census of 1951 listed 3,468 Nepalis in the erstwhile Lushai Hills. In the Naga Hills, they formed 4.29 per cent of the total population according to the 1961 census. During 1951–1971, the Nepalis/Gorkhas became a demographically significant group in Assam. They contested in electoral politics and were also elected to the Assam Legislative Assembly (Thapa 2003, 139–149). As per the 2011 Census of India, the number of Nepali-speaking Gorkhas in Assam number 596,210 only, or 1.91 per cent of the total population of Assam (31,205,576), which has the second largest concentration of Nepali-speaking population in India outside of Darjeeling district in West Bengal.[15]

The chapter shows that contrary to what different activists of the anti-foreigner movements believe, all Nepalis/Gorkhas settled in the Northeast are not foreign nationals but Indian citizens. Most of them settled in the region even before India became independent. Their presence and settlement in the region were the outcome of colonial and historical processes predating the Indian independence. Some Nepalis entered the region after independence, but there was nothing illegal about their entry, as their migration is legalized by the India–Nepal Friendship Treaty of 1950. Despite this, the Gorkhas have become targets of hostile attacks. There are several incidents of the dominant local majority targeting the Gorkhas/Nepalis settled in the region. In Assam, Manipur, Meghalaya, Mizoram and Nagaland, they have faced loss of property, eviction and displacement (Gurung 2002, 149–158). In the tribal belts and blocks and tribal autonomous councils of Assam, the Gorkhas face the problem of landownership because of lack of documents, right to primary education in their mother tongue, discrimination in employment and withdrawal of voting rights and grazing permits (Bhandari 2003, 122). To withstand the illegal and unjust onslaught, the Gorkhas/Nepalis are asserting their separate political identity as Indian Nepalis/Gorkhas as distinct from Nepali citizens of Nepal (Gurung 2016b, 233–256) and organizing into pressure groups to assert their right to live with dignity as Indian citizens.

NOTES

1. According to the 2011 census, the Nepali population in Northeast India total 1,247,461. Assam has 596,210 Nepali-speaking people, with Sikkim, Arunachal Pradesh, Manipur, Meghalaya, Nagaland, Mizoram and Agartala having Gorkha populations of 382,200, 95,317, 63,756, 54,716, 43,481, 8,994 and 2,787, respectively.

2. There is no ethnic group called Gorkha in Nepal. In Northeast India, the terms 'Gorkha' and 'Nepali' are used interchangeably to refer to Nepali speakers. The Indian Nepalis of the Northeast have consciously sought to identify themselves as Gorkhas to distinguish themselves from the Nepali

citizens of Nepal. The Nepali language, which was the sine qua non for the Gorkha soldiers in the colonial Indian army, emerged as the lingua franca of the Gorkha soldiers and of the Gorkhas settled in the North Eastern Region.

3. The modern political entity known as Nepal resulted from the unification of the country, then divided into a number of small states, by 1769 under Prithvi Narayan Shah, belonging to the hill kingdom of Gorkha.

4. As per the 2011 census of Nepal, 50 per cent of the total population is located in the Terai, 43 per cent in the hills and 6.73 per cent in the mountains. The major ethnic and caste groups within Nepal include the group called the Parbatiyas, whose mother tongue is Nepali (an Indo-Aryan language), constituting at present 40 per cent, the Newaris constituting 6 per cent, other hill communities or mountain-dwelling ethnic groups, such as the Magar, Gurung, Tamang, Rai, etc. who speak Tibeto-Burman languages, constituting 21 per cent, the Madhesis (speaking the North Indian languages Avadhi, Bhojpuri and Maithili) constituting 32 per cent and ethnic groups inhabiting the Inner Terai and the Terai constituting 9 per cent of the population.

5. Low marshy areas with deep, rich alluvium. The region grows most of the country's food. Half of the population of Nepal at present lives in the Terai plains lying to the south.

6. Under the Anglo-Chinese Convention of 1890, the Sikkim–Tibet border was agreed upon, and in 1895 it was jointly demarcated on the ground. Sikkim became a de jure protectorate of the British.

7. The first regular survey of Nepal was only conducted by the Survey of India in 1926–1927, which resulted in the actual demarcation of the Indo-Nepal boundary. There are, however, some boundary disputes that continue to exist between the two countries. From 1980 to 2007, a bilateral boundary working group headed by the respective surveyor generals of Nepal and India came together and decided to re-demarcate the boundary, for which a survey was conducted. About 98 per cent of the boundary demarcation has been settled, but political issues remain over the remaining 2 per cent.

8. During the British colonial period, the term 'Gurkhas' specifically referred to Nepali soldiers. The term 'Gorkha' at present refers to the Nepali-speaking community living in Darjeeling district of West Bengal and north-eastern states.

9. The term 'small wars' is used in the sense of smaller, punitive British expeditions carried out against hill tribes in Northeast India during the colonial period. It developed owing to the specific nature of the terrain, or ecology, and the tribal mode of warfare, which depended on ambushes and raids on enemy villages.

10. Report on the Administration of Assam for 1914–15. Shillong: Assam Secretariat.

11. *Census of India*. 1931. Vol III, Part 1-A-Report.

12. Foreign and Political Department confidential record. No.F.306-X/29, 11th February 1930: 5.

13. As per this agreement, wishes of the personnel of the regiments were sub-sequently conducted, after which four regiments of the Gurkha Rifles (2nd, 6th, 7th and 10th) remained with the British and six regiments (1st, 3rd, 4th, 5th, 8th and 9th) with India.

14. Every Nepali, by virtue of his/her domicile in the territory of India at the commencement of the Constitution of India (26 January 1950), under Article 5 of the Constitution, was a citizen of India (Gazette Notification on the Issue of Citizenship of Gorkhas, Government of India Ministry of Home Affairs, New Delhi, 23 August,1988).

15. As per the 2011 census, the total number of Nepali-speaking people in India was 2,926,168, constituting just 0.24 per cent of the total Indian population.

REFERENCES

Alavi, Seema. 1995. *The Sepoys and the Company: Tradition and Transition in Northern India 1770–1830*. Delhi: Oxford University Publications.

Bhandari, Purushottam L. 2003. 'Evolution and Growth of the Nepali Community in Northeast India'. In *The Nepalis in India: A Community in Search of Indian Identity*, edited by A. C. Sinha and T. B. Subba, 106–123. Delhi: Indus Publications.

Bhanskota, Purushottam. 1984. *The Gurkha Connection: A History of the Gurkha Recruitment in the British Indian Army*. Jaipur: Nirala.

Cohen, Stephen P. 2004. *The Indian Army: It's Contribution to the Development of a Nation*. New Delhi: Oxford University Press, Third Impression.

Government of India. 2008. *Annual Report 2007–08*. New Delhi: Department of Border Management, Ministry of Home Affairs.

Gurung, Tejimala. 2002. 'Displacing the Displaced: The Nepalis in Northeast India'. In *Dimensions of Displaced People in North-East India*, edited by C. Joshua Thomas, 149–158. New Delhi: Regency Publications.

———. 2008. 'Grazier Migration and the Rise of Dairy Industry in Assam'. In *Society and Economy in Assam*, Vol. 3, edited by D. R. Syiemlieh and Manorama Sharma, 159–169. New Delhi: Regency Publications.

———. 2009. 'Gorkhas as Colliers: Labour Recruitment and Racial Discourse in the Coal Mines of Assam'. In *Indian Nepalis: Issues and Perspectives*, edited by A. C. Sinha and T. B. Subba, 259–273. New Delhi: Concept Publishing Company.

———. 2015. 'The Making of Gurkhas as a "martial race" in Colonial India: Theory and Practice', in *Proceedings of the Indian History Congress*, Aligarh, pp. 520–529.

———. 2016a. 'Gurkha Displacement from Burma in 1942: A Historical Narrative'. In *Nepali Diaspora in a Globalised Era*, edited by Tanka B. Subba and A. C. Sinha, 203–220. London: Routledge.

Gurung, Tejimala. 2016b. 'Autonomy Demand of the Gorkhas in Assam: A Preliminary Note'. In *Ethnicity and Political Economy in Northeast India*, edited by H. Srikanth and Rooplekha Borgohain, 233–256. Guwahati: DVS Publishers.

———. 2016c. 'Role of the 8th Gurkha Rifles in British Expansion in North-East India: A Preliminary Note.' In Proceedings of NEIHA, Thirty Seventh Session, Sikkim.

———. 2017. 'Rethinking Nepali Migration in Colonial Assam: A Gender Perspective', in *Rethinking Gender History: Essays on Northeast India and Beyond*, edited by Manorama Sharma, 216-238. New Delhi: D.V.S. Publishers.

Huxford, H. J. 1952. *History of the 8th Gurkha Rifles 1824–1949*. Aldershot: Gale and Polden.

Lama, Mahendra P., and Vimal Khawas. 2009. *Problems on Border Areas in North East India: Cases from Darjeeling and Sikkim Himalaya*. Gangtok: Sikkim University. https://www.academia.edu/34256281/Problems_on_Border_Areas_in_North_East_India_Cases_from_Darjeeling_and_Sikkim_Himalaya (accessed 25 November 2020).

Mojumdar, Kanchanmoy. 1975. *Nepal and the Indian Nationalist Movement*. Calcutta: Firma K. L. Mukhopadhyay.

Muni, S. D. 1973. *Foreign Policy of Nepal*. Delhi: National Publishing House.

Omissi, David. 1994. *The Sepoy and the Raj: The Indian Army, 1860–1940*. Basingstoke: Macmillan.

Shakespear, L. W. 1977. *History of the Assam Rifles*. Calcutta: Firma K.L.M.

Thapa, R. 2003. 'Nepali Participation in the Electoral Politics of Assam'. In *The Nepalis in India: A Community in Search of Indian Identity*, edited by A. C. Sinha and T. B. Subba, 250–261. Delhi: Indus Publications.

Waterman, David M., ed. 2004. *The Origins of Himalayan Studies: Brian Houghton Hodgson in Nepal and Darjeeling 1820–1858*. London: Routledge.

Whelpton, John. 2010, reprint. *A History of Nepal*. Noida: Cambridge University Press.

Chapter 13

Burmese Indians
Growth of Burmese Nationalism and Ethnic Discrimination

Emdorini Thangkhiew

INTRODUCTION

The term Burmese Indians refers to the community of Indian-origin ethnic groups that migrated to Burma, that is, present-day Myanmar, when Burma was part of British India. While most of the migrants were forced to return to India after Burma became independent, there are still some Indian-origin Burmese Indians in present-day Myanmar.[1] According to the Singhvi Committee (2009) report, in Myanmar, there are still 2.9 million people of Indian origin who account for 2 per cent of the total population (Pai 2017). Some believe that the total people of Indian origin might even constitute 4 per cent of the total population or more (Chaturvedi 2017). These Burmese Indians, who are the third or the fourth generation of the Indian-origin migrants, have never seen India. The present chapter throws light on the history of the Indian migrants, their accomplishments and their travails and experiences in an alien land where the natives envied and refused to acknowledge them as Burmese and marginalized their presence and role in Burma/ Myanmar. It focuses on how the ethnic conflicts and clashes between the Indian-origin migrants and the native Burmese led to the assertion of Burmese nationalism with Buddhist orientation, eventually leading to the discriminatory practices that compelled many migrants of Indian origin to return to India. The chapter also throws light on the status of the Burmese Indians who continue to stay in Myanmar and the consequences of living amid 'popular hostility and patronizing attitudes from the rest of the Buddhist-dominated Burmese society' (Egreteau 2011). It also examines the efforts made by the Indian government to address the problems of the Indian–origin community in Burma, and what more it should do to improve the 'fairly impoverished' people of Indian origin in Myanmar.

HISTORY OF INDIAN MIGRANTS IN BURMA

To understand the causes of the anti–Indian sentiments in Burma, we need to look into colonial history. The British began acquiring a hold over Burma after their success in the First Anglo–Burmese War in 1824. After the Anglo–Burmese War of 1885, the whole of Burma came under British control, and it became a province of British India. The British took along with them the Indians familiar with colonial administration to this unfamiliar territory to assist them in realizing the colonial goals. They required hardworking and skilled labourers to ensure the exploitation of natural resources and consolidate their colonial economic and political interests in Burma. As the native Burmese labourers were unskilled and inadequate, the British, who were in favour of interdependent trade and labour movements throughout their colonies in South Asia encouraged large-scale migration of skilled Indians into Burma. Since India and Burma were connected by land and water, it was easier to move the Indian workers permanently to Burma. The labourers from the distant Arad, Dumraon and Faizabad in Bihar, Uttar Pradesh, Punjab and Bengal were relocated to Burma. Many migrant labourers came from the southern part of British India, from the Tamil- and Telugu-speaking areas of the Madras Presidency (Satyanarayana 2010). A steady stream of Indians moved to Burma as civil servants, traders, engineers, river pilots, farmers, indentured labourers and artisans, soldiers and medical practitioners. The inflow of Indians increased the percentage of Indians in Burma through the years. In 1872, the percentage of the Indian population was recorded at 4.9 per cent, and by the year 1931, it increased to 6.9 per cent of the population (Baxter 1941).

Soon, the people of Indian origin became the backbone of the economy. Many of them settled in Lower Burma and were involved in rice production. During British rule, Burma emerged as the largest exporter of rice. The growth in rice production led to the development of other economic activities. The people of Indian origin and the people of Chinese origin with their vast knowledge of rice cultivation contributed to the growth of the Burmese economy. The Indian labourers were engaged in all spheres of agricultural activities, from bunding to threshing and from shipping to road transport. The Chettiars from South India became primarily moneylenders for the farmers and peasants. They contributed immensely to agrarian growth in Burma during the colonial era (Turnell and Vicary 2008). The British took charge of exporting the rice globally, and this led to the rise of the port towns like Rangoon and Sittwe. The labourers of Indian origin hired in the rice processing mills lived in most of the port towns (Cheng 1968). They were willing to work for far lower wages than the local Burmese (Satyanarayana 2010). Some Indians who made it big became monopoles in certain sectors, such as automobile, electrical goods, education and money lending. Under British control and supervision, they started dominating the local economy, military sector, administrative sector, education, health, etc. The early Indian settlers, through their gains in the economy, educated their children and occupied top positions in universities and bureaucracy.

The prominent positions that the Indians had in the British administration and the financial gains they achieved in Burma slowly gave way to anti-Indian sentiments (Pai 2017) and resentment among the native Burmese people (Cowan 1975). The Great Depression led to intense competition between the people of Indian origin and the Burmese natives, who were then willing to perform even substandard menial jobs. The peasants and farmers were hit hard by the depression, as the price of rice plummeted. The Chettiars exploited the peasants and confiscated the land and livestock of the native Burmese peasants who could not pay back the loans. Their role as moneylenders made the Chettiars a target of the Burmese hatred (Egreteau 2011).

Soon, the native ire turned against the Indian workers. In Rangoon, the Indian workers went on a strike against the port authorities to have their demands met. In May 1930, Stevedores, the British firm operating in the port, engaged Burmese workers to replace the Indians labourers on strike. Unable to get their demands met, the Indian labourers resumed their work. When the Burmese came to report for work the next morning, they were informed that their services were not needed. This angered the Burmese natives, and the fight between the two groups of workers sped up into an anti-Indian riot (Tun 1938). At least 200 Indians were killed and flung into the river in the first hour. The police were called in and Section 144 of the Criminal Procedure Code was invoked, prohibiting an assembly of more than four or five people. The natives' unrest eventually spread across the entire country. The 'Burma for Burmese only' campaign was launched, which took a violent turn, leading to the looting of shops, burning of houses and destruction of places of worship. The Burmese nationalists began demanding freedom from not only the British domination but also the Indian economic hold (Suryanaraya 2009).

The Japanese occupation of Burma in 1942 worsened the economic conditions of the Burmese Indians. It destroyed the economy and adversely affected the position that the Indians had so far enjoyed. Over 5 lakh Burmese Indians were forced to flee the country, and almost half of them failed to make it to their destination (Suryanaraya 2009). This massive exodus led to the decline of the Indian population (Mahajani 1960), which would never increase because of the waves of migration outside Burma in the years of the Japanese invasion and during General Ne Win's military rule. Following the Japanese occupation, the weakened Burmese Indians witnessed a loss of position and prestige in the Burmese society, from which they could never rise again.

NATIONALIST POLICIES AND THEIR CONSEQUENCES

While granting independence to Burma in 1948, the British made no efforts to secure citizenship for the people they brought from India. The newly independent government, trying to assist its people, redistributed land to the native peasants. The land reforms implemented by the government took away a vast tract of irrigated land from the Tamil Indians in Burma. The Burmese government initiated many policies that served the interests of the Burmese, such as the Standard Rent Act, Tenancy Disposal Act, Agricultural Debt Relief Act, Land Nationalization Act, Agricultural Bank Act and Burma Foreigners Act (Suryanaraya 2009). The Chettiars bore the brunt of these policies more than anyone else. They were deprived of their wealth, jobs and property. Compared to the loss that they incurred, the compensation they received from the government was very meagre. Stringent foreign exchange rules passed by the government made it difficult for them to repatriate money to India. Further, the government imposed Burmese as the official language and medium of instruction in all schools. All Tamil schools were closed. Teaching of Hindi was not allowed in schools. In all government jobs, the Burmese were given preferential treatment.[2] The new recruitment policy affected many people of the Indian origin who were employed earlier in government departments and public sector organizations. The official preference to the language, culture and religion of the Burmese went against the interests of the smaller minority groups. Although the native ethnic minorities were accommodated to some extent, the Indian and Chinese communities that had migrated to Burma centuries ago lost their status as ethnic minorities and were labelled as aliens (Egreteau 2011).

Following the anti-Indian riots of 1930, the Japanese invasion and the Burmanization policy of the government, the numerical strength of Indian-origin people gradually declined. The situation worsened after General Ne Win seized power through a military coup in 1962. General Ne Win, in the name of nationalism, banished anyone or anything that he considered as not compatible with Burmese culture, be it cricket (a favourite sport of the Burmese Indians), beauty contests or even the teaching of English in the universities. During his regime, more than 3 lakh Burmese Indians were forced out of Burma. The poor migrants who stayed back became the target of the regime's brutality. The junta did not spare even the third and fourth generations of Indian families that had been living in Burma for centuries and gotten assimilated into Burmese society. Those who were left behind were reduced to the status of an alien minority. Only a few among them were granted citizenship. The Burmese Indians who had been relatively well off during the colonial regime became one of the marginalized sections of the society (Scott 1882). Their privileges and freedom were curtailed (Schober 2006). Ne Win's regime imposed restrictions even on the right to freedom of religion guaranteed under the constitution. The period of 'Forced Burmanization' led to the impoverishment of the Indian community (Devi 2014).

In pursuit of the official policy of 'Burmese Way to Socialism', over 15,000 enterprises were taken over by the government between 1963 and 1972. The properties of thousands of Indians were seized. The government nationalized even small business enterprises, banks and warehouses and denied trading licenses and government jobs to the people identified as non-Burmese. The non-Burmese ethnic minorities and other minorities, such as Indians, Chinese, Anglo-Burmese and Westerners were the prime targets of the radical nationalization policy of the Burmese government (Holmes 1967). As Chaturvedi put it, 'Economy was overhauled, Indian and Chinese businesses and trade contracts were rescinded, and the Indian-majority Cabinet was dissolved' (Chaturvedi 2017). The wholesale and retail businesses of the people of Indian origin were taken away, and they were given a meagre 175 Kyat to return to India (*The Irrawaddy* 1999). Those fleeing the country were not allowed to carry back any of their savings. Even their valuables and properties were confiscated. It was claimed that women were not even allowed to take back with them their *mangalsutra* (a necklace that represents a married lady in the Hindu religion). The Indian government arranged ferries and aircraft to lift the Burmese Indians who were forced out of Burma.

On the eve of independence, both India and Burma had a working relationship with each other. They even signed a Treaty of Peace and Friendship in 1951. However, the relations between the two soured during Ne Win's regime. They witnessed a downfall as India strongly opposed the imposition of the military regime. The relentless support India had given to the pro-democratic forces drove a wedge in its relations with Burma. Ne Win's regime withdrew Burma from the Non-Aligned Movement in 1979 (Kanwal 2010).

CITIZENSHIP RULES

After gaining independence in 1948, the newly independent government of Burma introduced citizenship regulations. The new constitution made it a condition that to be eligible for citizenship one should have been a resident in Burma continuously for at least 8 years preceding the Second World War. The conditions for citizenship became strict after Ne Win took over political power. Because of the instability and political uncertainty that followed the assassination of Aung San, many Indians did not apply for citizenship. The applicants were asked to give concrete evidence that their ancestors had settled in Burma after the First Anglo-Burmese War (1824–1826). The application process was so complicated that out of the 400,000 who applied, only 10,000 were granted citizenship.

According to the citizenship law passed in 1982, Burmese were divided into three categories of citizens: full citizens, associate citizens and naturalized citizens. Full citizenship was granted only to those who could furnish 'conclusive evidence' of entry and residence in Myanmar before the British annexation and to those who were proficient in one of the national languages and whose children were born in the country. These criteria created a problem for many of the minorities, for even though they met the eligibility requirements, they could not support their claim with any documentary proof to show that their families had been living in Burma before 1823 (International Labour Organization 1982). Associate citizenship was granted only to those who applied for citizenship under the Union Citizenship Act 1948. Naturalized citizens referred to those who had been living in Burma before 4 January 1948 and applied for citizenship after 1982.

One hundred and thirty-five national races were granted full citizenship. Because ethnic Indians were not considered national, they were excluded from full citizenship. However, they were allowed to apply for the two lower tiers of citizenship with limited rights. Burmese law treated a large percentage of the Indian community as 'resident aliens'. Though many of them had long ties with Burma or were born there, they were not considered citizens under the 1982 Burma Citizenship Law that restricted citizenship for groups that immigrated before 1823.

Even though the law of 1982 established a path for Indians to citizen-ship, it had a strict racial definition of citizenship. The International Commission of Jurists (ICJ) declares the Burma Citizenship Law of 1982 as incompatible with international human rights law, as it violates the Universal Declaration of Human Rights, the Convention on the Rights of the Child and international norms that prohibit discrimina-tion against racial and religious minorities.

Because of the stringent laws for citizenship, a vast number of people failed to attain citizenship. They have become 'stateless' people having no access to education, healthcare, employment and travel documents. The biggest problem for Burmese Indians is their inability to get the national registration card (NRC). The NRC is an in-demand identity card granted to Burmese citizens. It is indispensable, as it is needed for all public and official transactions. It is needed even for buying a train ticket (Han 2016).

CULTURAL AND IDENTITY CHALLENGES

Myanmar has a democratic system today, but anti-Indian prejudice still prevails in the social and political culture of the predominantly Bamar-dominated Burmese society. Burmese culture has always been xenophobic and looked down upon others of a different ethnicity, race, religion or skin colour (Rydgren 2004). Unlike the Chinese, who got completely integrated, the Indian community showed an unwillingness to get absorbed into Buddhist culture and kept their faith and customs. As the Indians were dark-skinned and practised mostly Hinduism or Islam, intermingling with the locals was difficult for them. Indians were discriminated against because of their skin colour, appearance and faith.

Although the locals looked at the Indian community as a homoge-nous community, the Burmese Indians belonged to different religions and regions. They included Hindus, Muslims, Sikhs and Christians. They had migrated from different parts of British India, which was partitioned into India, Pakistan and Bangladesh. They spoke such languages as Hindi, Punjabi, Telugu, Malayalam, Marwari, Gujarati, Parsi, etc. They were economically stratified. Alongside the rich Chettiars, who were envied, there were also poor Tamils/Bengalis who performed menial jobs, such as rickshaw pulling, sweeping, handling of dead bodies, etc., who were despised, and also educated Indians who worked as doctors, lawyers and teachers. However, once the Indophobia started taking shape, these differences did not matter to the locals or the state. Everyone was despised as outsiders (Mahajani 1960).

Even though the 'Indophobia' started in the 19th century, it manifested physically in the anti-Indian riots of 1930 and thereafter continued throughout the 20th century. It becomes apparent in the policies of the independent government, in the orders passed during the Ne Win regime, in the cultural practices and in the linguistic and economic activities. The military dictatorship used religion as a tool for oppression (Hre 2013). Restrictions were imposed on the celebration of Hindu and Muslim festivals. The government expected the Burmese Indians not to create any problem or inconvenience for the natives while celebrating their festivals. Any request for permission to build a mosque, or to go abroad for a religious purpose, was summarily rejected. The government introduced Burmese as a compulsory language for all minority communities. Indophobic expressions were articulated in the day-to-day Burmese language. The Indian community were derided as *kala lumyo* (dark-skinned) (World Heritage Encyclopedia 'Expulsion of Indians'), *firangi* (foreigner) or *chetti-kala* (other). The situation has indeed been becoming better in recent decades. The Hindus, Sikhs and Christians are facing fewer hostilities. Even though they are still considered as outsiders, the Hindus are now allowed to practise their faith. In contrast, as Butkaew points out, 'the Muslims continue to experience the most severe forms of legal, economic, religious, educational, and social restrictions and discrimination' (Butkaew 2005). There is a widespread irrational fear of the spread of Islam and the Islamization of the Burmese society. Many of the Burmese-Indian Muslims have settled in Rakhine State, or Arakan State, which is geographically isolated from mainland Myanmar. Separated by the Bay of Bengal, this state is closer to Chittagong in Bangladesh. Under General Ne Win, the Tatmadaw (armed forces of Myanmar) expelled many of the Arakan Rohingya Muslims in an operation code-named 'Operation King Dragon'. In 1990, the military junta renamed Arakan State as Rakhine State because of the dominant Rakhine ethnic group. The Burmese

Muslims, especially the Rohingya minority, became the latest target of the Burmese government (Cowan 1975).

Apart from legal hurdles, Indians are also faced with administrative hurdles that make it very difficult for them to rise to the top ranks in any sector, be it a chief executive officer (CEO) in a company or a colonel in the military. In the governmental departments, they cannot rise above the level of a director. Indians are also prevented from advancing as civil servants or working in companies run by state governments. Burmese leaders are worried about slipping into 'foreign hands'. They favour only Burmese businessmen in all sectors of the economy.

POLITICAL ECONOMY OF BURMESE-INDIAN SETTLERS AND MIGRANTS

Of the 2.5–3 million Burmese Indians still in Burma (2010–2011), the majority of them are living in the urban cities of Yangon, Mandalay and Ziyawadi, port towns of Moulmein, Pathein and Sittwe and old colonial towns of Pyin U Lwin and Kalaw. They are involved in small businesses, such as running of restaurants, jewellery stores, money exchange trading, pharmacies, etc. Some, like the Burmese Tamils, turned to illegal trade in drugs and forest products along the India–Burma borders. Although many do not have any links with their homeland in India, one can still see many practising the customs and language of their native country. However, there are others who have been completely integrated into their new homeland and have forgotten the language and customs of their forefathers. They have Burmese names and Burmese identity cards and are more at ease with the *bamazaga*, the national language of Burma, than with Hindi or Bengali. They have shown love for their new homes, lost all contact with their original homes and are disinterested in ever returning to India (Han 2016).

Of the many that have fled Burma, some have settled in Tamil Nadu, and others have settled in the four north-eastern states that share borders with Myanmar, namely Nagaland, Manipur, Mizoram and Arunachal Pradesh. Some have settled as traders in Moreh, a small Indian town on the India–Myanmar border in Manipur. Since they know the Burmese language, they are able to trade with the Burmese who come to Moreh in search of such goods as automobile parts, clothing and cosmetics. One can also see settlements of Burmese Indians in Subhasgram East near Kolkata, which is now called 'Burma Colony', Madras (Chennai) and around the harbour of Visakhapatnam in Andhra Pradesh.

CONCLUSION: INDIA'S FAILURE

India did little to help the people of Indian origin living in Burma. The Indian government no doubt took measures to rehabilitate Burmese Indians who returned to India. However, it hardly made any effort to influence the government in Burma (later Myanmar) to make the living conditions of the people living there better. On the eve of independence, India and Burma concluded their Friendship Treaty and concluded agreements on the Indo-Myanmar border. Later, Burma became a non-aligned country. During this period, India could have used its friendship to influence the Burmese government to be fair to the Indian community. When the military took over political power in Burma, India, like China, could have continued to maintain its relationship with Burma through strengthening bilateral relations. However, India's stand made Burma distance itself from India and look towards China. China used its growing military and economic strength to improve its relationship with Myanmar and influenced the government in Myanmar to recognize the rights of the people of Chinese origin and treat them as equal citizens (Farrelly and Olinga-Shannon 2015). However, this was something that India failed to do. India could have used its historical linkages, its Buddhist heritage and its long border with Myanmar as opportunities for building good relations with its eastern neighbour. The Burmese Indians could have become a link in strengthening the bonds between the two. The Indian government could have used the diaspora as a 'strategic asset' to strengthen its bilateral links with Myanmar (Kapur 2003), but it failed to take advantage of such available opportunities. It was only in recent years that India started rebuilding its relations with Myanmar (Egreteau 2003), but whether it would be of any consequence to Burmese Indians or not is yet to be seen.

NOTES

1. In 1989, the military government changed the name of Burma to Myanmar. India accepted the name change. Hence, in this chapter, we use the term Burma up to 1989 and Myanmar thereafter.
2. Burma State Department. 2017. Available at https://www.state.gov/documents/organization/160450.pdf (accessed on 1 May 2018).

REFERENCES

Baxter, J. 1941. *Report on Indian Immigration*. Rangoon: Government Printers.

Butkaew, Samart. 2005. 'Burmese Indians: The Forgotten Lives'. *Burma Issues: Peace Way Foundation* 16 (2): 1–3.

Chaturvedi, Medha. 2017. *Indian Migrants in Myanmar: Emerging Trends and Challenges*. India Centre for Migration. New Delhi: Ministry of Overseas Indian Affairs.

Cheng, S. H. 1968. *The Rice Industry of Burma: 1852–1940*. Kuala Lumpur: University of Malaya Press.

Cowan, C. D. 1975. 'Book Review of Moshe Yegar: The Muslims of Burma: A Study of a Minority Group'. *Bulletin of the School of Oriental and African Studies* 38(2):486–487.

Devi, Konsam Shakila. 2014. 'Myanmar Under the Military Rule 1962–1988'. *International Research Journal of Social Sciences* 3(10):46–50.

Egreteau, Reanand. 2003. *India and Burma/Myanmar Relations: From Idealism to Realism*. Conference. New Delhi: India International Center.

Egreteau, Renaud. 2011. 'Burmese Indians in Contemporary Burma: Heritage, Influence, and Perceptions Since 1988'. *Asian Ethnicity* 12 (1): 33–54.

Farrelly, Nicholas, and Stephanie Olinga-Shannon. 2015. *Trends in Southeast Asia; Establishing Contemporary Chinese Life in Myanmar*. Singapore: ISEAS Publishing.

Han, Thi Ri. 2016. 'Myanmar's Hindu Community Looks West'. *Frontier*. https://frontiermyanmar.net/en/myanmars-hindu-community-looks-west (accessed 10 March 2018).

Holmes, Robert. 1967. 'Burmese Domestic Policy: The Politics of Burmanization'. *Asian Survey* 7(3):188–197.

Hre, Mang. 2013. 'Religion: A Tool of Dictators to Cleanse Ethnic Minority in Myanmar?' *IAFOR Journal of Ethics Religion & Philosophy* 1(1):21–29.

International Labour Organization. 1982. 'Burma Citizenship Law'. http://www.ilo.org/dyn/natlex/natlex4.detail?p_lang=en&p_isn=87413&p_country=MMR&p_count=86 (accessed 5 March 2018).

Kanwal, Gurmeet. 2010. 'A Strategic Perspective on India-Myanmar Relations'. Research Gate. https://www.researchgate.net/publication/292667167_A_strategic_perspective_on_India-Myanmar_relations (accessed 10 March 2018).

Kapur, Devesh. 2003. 'Indian Diaspora as a Strategic Asset'. *Economic and Political Weekly*, 38(5): 445–448.

Mahajani, Usha. 1960. 'The Role of Indian Minorities in Burma and Malaya'. *Journal of Southeast Asian History* 3(1):143–147.

Pai, Nitin. 2017. 'Why India Must Be Tough with Myanmar'. *Rediff News*, December 24. http://www.rediff.com/news/column/why-india-must-be-tough-with-myanmar/20171224.html (accessed 21 May 2018).

Rydgren, Jens. 2004. 'The Logic of Xenophobia'. *Sage Journal* 16(2):123–148.

Satyanarayana, Adapa. 2010. ' "Birds of Passage": Migration of South Indian Laborers to South East Asia'. *Critical Asian Studies* 34(1):89–115.

Schober, Juliane. 2006. 'Buddhism, Violence, and the State in Burma (Myanmar) and Sri Lanka'. In *Religion and Conflict in South and Southeast Asia: Disrupting Violence*, edited by Linell E. Cady and Sheldon W. Simon, 59–61. London: Routledge.

Scott, James George. 1882. *The Burman—His Life and Notions*. New York: The Norton Library.

Singhvi Committee. 2009. *World Affairs, Subcontinent Central Asia (Report)*. Singhvi Committee.

Suryanaraya, V. 2009. *The Indian Community in Myanmar* (Paper No. 3523). South Asian Analysis Group.

The Irrawaddy. 1999. 'Indian and Burma: Working on Their Relationship'. *The Irrawaddy* 7(3). https://http://www2.irrawaddy.com/article.php?art_id=1170 (accessed 5 March 2018).

Tun, Than 1938. 'Race Riots in Burma'. *Workers' International News* 1(9):8–10.

Turnell, Sean, and A. Vicary. 2008. 'Parching the Land? The Chettiars in Burma'. *Australian Economic History Review* 48(1):1–25.

World Heritage Encyclopedia. 'Expulsion of Indians From Burma in 1962'. http://self.gutenberg.org/articles/eng/expulsion_of_indians_from_burma_in_1962 (accessed 15 May 2018).

Chapter 14

Returnees and Refugees from Burma to India
Differing State Responses

Saurabh Kaushik

> Burmese Chin refugees, settled in Delhi for years now, protested out-
> side the building of the United Nations Commissioner for Refugees
> (UNHCR) today against what they described as hideous living condi-
> tions, dearth of jobs and lack of necessary treatment for the seriously ill.

This was what R. Nivedita, a correspondent with *The Statesman*, an
Indian English daily, reported on 19 June 2015 in her article titled,
'Burmese Chin Refugees Protest at UNHCR' (Nivedita 2015). The
Burmese protest was not the first of its kind. For several years, Burmese
have been protesting at United Nations High Commissioner for
Refugees (UNHCR) premises and have faced police repression, tor-
ture and even threats of being repatriated to Burma (Chin Refugee
Committee 2011). Another article titled, 'Islamic countries Shut Door
but India welcomed Us: Rohingyas', published in the *Times of India*
a month after, threw light on the condition of about 105 Muslim
Rohingya families who have been put up at Welcome Colony and
Mehnat Nagar in the Hasanpura area of Jaipur (Khani 2015). The quest
of these refugees fleeing persecution and state repression for eking out

a survival in various parts of India and being legitimately recognized as refugees is fraught with several hardships (Ghoshal 2013).

Re-migration (implying the return of Indian expatriates and migrant labourers who had settled down in Burma during the British rule) and flow of refugees from Burma into India is a phenomenon that dates back to the 1930s when Indian migrants who had ventured into Burma to seek economic opportunities in the late 19th century began crossing the international border between the two countries. During the 1930s, owing to rising Burmese nationalism, there was a surge in anti-Indian sentiment engendered by a lack of assimilation of Indians in the Burmese society. The dire economic impact of the Great Depression particularly on native Burmese led to further resentment against the Indians which ultimately resulted in the flow of 'returnees' into India (Egreteau 2013). It is important to distinguish between the returnees who came to India from Burma and the migration of Chin and Rohingya refugees. While the returnees were ethnic Indians hailing mainly from Bihar, Odisha, Andhra Pradesh and Tamil Nadu, the refugees include the Chin, who are legitimate Burmese citizens, or the Rohingya, who were living in Burma for many generations but were denied citizenship rights.

There was a substantial flow of Indian communities by sea, by air or through the jungles of Burma into India after the Japanese invasion of Burma in 1942 during the Second World War. After Burma became independent, the outflow of Indians from Rangoon, Akyab and other regions of Burma into India increased. This process intensified after General Ne Win seized power in a coup in 1962 and passed a series of laws and enacted policies resulting in further nationalization and 'Burmanization' of the state. Those leaving were mostly Tamils (among whom the Chettiars had been influential landowners and moneylenders during the colonial period), but there were also Punjabis, Biharis, Bengalis, Telugus and Marwaris who had to flee the land that was a source of their lives and livelihood (Mehrotra and Basistha 2011). Many such returnees, who played important roles as traders, merchants, clerks, teachers, civil servants and farm labourers and contributed to the social and economic development of the host country, felt uprooted and suffered a tremendous sense of loss when they were forced to leave the country.

The state repression and the crackdown on pro-democracy students and protesters during the 1980s and afterwards led to many students and political activists seeking refuge in the bordering Manipur state of India (Bobichand 2015). Around the same time, the state perpetrated ethnic discrimination and religious repression against ethnic minorities, including the Chin. The Chin had to endure 'extrajudicial killings, arbitrary arrest and detention, torture and mistreatment, forced labor, severe reprisals against members of the opposition, restrictions on movement, expression, and religious freedom, abusive military conscription policies, and extortion and confiscation of property' (Human Rights Watch 2009a). With the onset of the military dictatorship in 1978 and the enactment of the Burmese-nationality law in 1982, religious and racial hatred against the Rohingya Muslims accentuated. The state actions led to persecution and sectarian killings, forcing them to seek refuge in neighbouring countries, including Bangladesh and India. Since then, ethnic violence has erupted intermittently and has adversely affected the Rohingya (International Crisis Group 2014). The intensity of persecution suffered by the Rohingya compelled the British Broadcasting Corporation (BBC) to declare them 'among the world's least wanted' and 'among the world's most persecuted minorities' (Dummett 2010).

This chapter has three segments. The first contains an exposition of the causes and patterns of the influx of three categories of migrants from Burma to India: (a) the Indian-origin returnees; (b) the Chin; and (c) the Rohingya. Subsequently, in the other two segments, the chapter attempts to compare and analyse two interrelated aspects of re-migration and refugee flows from Burma into India and examines the varied responses of the Indian government to these different categories of people who migrated to India. While analysing these aspects, the chapter sheds light on how an intentionally selective reading of both national and international legal frameworks informs the actions of the Indian state towards migration and refugee influx from Burma. After addressing the aspects highlighted above, it concludes with a few recommendations regarding making India's refugee policy clearer, consistent and more in tune with international humanitarian law and international refugee law.

CAUSES AND PATTERNS OF INFLUX INTO INDIA

There are varied and complex causes to the influx of people from Burma into India. These causes, besides the patterns of the influx, are analysed below for the three ethnic groups, namely ethnic Indian 'returnees', the Chin and the Rohingya.

THE FLIGHT OF THE 'RETURNEE' INDIANS: 'BURMANIZATION' AND SECOND WORLD WAR

Beginning in the 1890s and over five decades after the British occupation of Burma in 1824, Indians were the single largest group in Rangoon, as highlighted in Table 14.1. Over 55 per cent of the tax revenue collected in Rangoon in 1931 came from Indians, and of the 1,567,315 inland and foreign telegrams sent in and out of Burma in 1938–1939, 670,959 belonged to Indians. In the beginning of the Second World War II (1939), Indians formed 16 per cent of the total population of Burma (Bhaumik 2003).

According to an estimate, there were over 1 million persons of Indian origin in Burma in 1931 working in different fields. Despite their contributions to the economy, the Burmese people envied the Indian migrants and never viewed them as Burmese citizens. The Burmese, who made up about 68 per cent of the population of the central plains,

Table 14.1 *Indians in Rangoon*

Year	Rangoon's Total Population	Indians in Rangoon	Percentage of Indians to Total Population
1891	180,324	87,487	49.5
1901	248,060	119,290	48.1
1911	243,316	165,495	56.5
1921	341,962	187,975	55.0
1931	400,415	212,692	52.9

Source: Bhaumik (2003).

found the rising population and growing economic power of Indians in Burma to be antithetical to their interests. With the growth of Burmese nationalism, the locals started attacking the Indian migrants and forcing them to return to India. The first ones among the ethnic Indians who returned to Indian territory during the 1930s and 1940s were only treated as 'returnees' fleeing persecution and were not considered refugees. Their journey was largely perceived to be a homecoming, albeit one that was riddled with difficulties. The situation became grim once the British colonial rule ended. Following the independence of Burma on 4 January 1948, the Burmese government adopted a series of legislations that affected the Indians either directly or indirectly. In the name of 'Burmese Way to Socialism', the government pursued three strategies: nationalization, industrialization and 'Burmanization'. These laws included, among others, the Land Alienation Act, 1948 and Land Nationalization Act, 1949, which led to elimination of absentee landowners and moneylenders (primarily targeting the Tamil Chettiars) and redistribution of land among landless Burmese peasants. The Industrial Regulation of 1948 ensured that Burmese had to own a majority share in all private enterprises (including small-scale retail trade). The Citizenship Act of 1982 denied citizenship to the migrant people of Indian origin and started seizing their land or property. Denial of citizenship and exclusion from public services, trade and business activities compelled many Burmese of Indian origin to come back to India (Bhaumik 2003).

The xenophobia that gripped Burma from 1962 onwards after the seizure of power by General Ne Win further resulted in the efforts at re-indigenization (or 'Burmanization') of resources through enactment of such laws as the Enterprises Nationalization Act of February 1963. The policies of Ne Win resulted in an unprecedented influx of Indians from Burma (Egreteau 2013). By the end of 1987, when the Indian government forced the Burmese government to halt the repatriation of Burmese Indians, only a small proportion of ethnic Indians remained in Burma. Indians today constitute a mere 2 per cent of the population. Several Indian repatriates from Burma were resettled in various parts of eastern India, Tamil Nadu, Andhra Pradesh, Orissa, West Bengal and suburbs of Delhi. The attempt was primarily to resettle the returnees largely in areas they inhabited in India before migrating to Burma.

Of these, many repatriates were resettled in 'peri–urban' areas, because a majority of them had set up their lives in and around urban centres, particularly in Rangoon, during their stay in Burma. However, there were the Burmese Bihari, Oriya and Andhra farmers, who had settled in the rural districts of Zeyawadi and Kyauktaga in Central Burma, that formed the exception, as they were resettled in rural lands in Andhra Pradesh, Bihar (near Patna), Madhya Pradesh (Bhopal and Betul districts) and northern Uttar Pradesh, near Rudrapur and Hastinapur, to be provided with an agricultural life that they had practised in Burmese villages. For a detailed overview of the transit camps, peri–urban and rural settlements of the returnees from Burma, see Table 14.2.

Table 14.2 *'Burma Colonies' in India: Select Urban and Peri-urban Resettlements of Burmese-Indian Repatriates Since the 1960s*

Type of Settlement		Places of Resettlement
Type 1	Main transit camps and evacuee centres for repatriates	1. Bettiah (Bihar) 2. Kancharapalem (Andhra Pradesh) 3. Gummidipoondi (Tamil Nadu)
Type 2	Sites of significant urban and peri-urban Burma colonies	1. Rudrapur, Hastinapur (UP) 2. Ranhola, Vikas Puri (Delhi) 3. Barasat, Kamarhati Camp, Subharanga (West Bengal) 4. Vyasarpadi, Periyar Nagar (Tamil Nadu) 5. Samastipur (Bihar) 6. Betul (Madhya Pradesh)
Type 3	Sites with major urban Burma 'colonies' and 'bazaars'	1. New Delhi (NCT of Delhi) 2. Jaipur (Rajasthan) 3. Patna (Bihar) 4. Bhopal (Madhya Pradesh) 5. Kolkata (West Bengal) 6. Moreh (Manipur) 7. Visakhapatnam (Andhra Pradesh) 8. Vellore, Trichy, Madurai and Chennai (Tamil Nadu)

Source: Egreteau (2013).

THE FLEEING CHIN: A CASE OF BETRAYED ASPIRATIONS AND MILITARY OPPRESSION

For centuries, the Chin have regulated their affairs as autonomous tribes or chiefdoms unencumbered by the travails of the modern nation state. The attempts by the British from 1872 to 1879 to invade Chin territory through present-day Bangladesh in the east, India's Assam state in the north and Burma in the east resulted in British control over large parts of Chin land, who then divided it into three administrative units, roughly dividing them between the modern-day nation states of India, Burma and Bangladesh (Human Rights Watch 2009a).

Although the Chin were part of the 'Panglong Agreement' of 12 February 1947 which guaranteed a federal union and an autonomous province for the Chin, the assassination of General Aung San (Burma's chargé d'affaires in negotiating independence with the British and crafter of the agreement) right before independence meant that Chin aspirations remained unfulfilled. The oppressive military rule in Burma under General Ne Win culminated in the 8-8-88 Uprisings in August 1988 which resulted in a brutal crackdown on protesters by the Tatmadaw, resulting in 3,000 deaths. Infantry Battalion number 89 (IB 89) from Kalaymyo suppressed the Chin students who protested. In fact, before 1988, the Tatmadaw had no battalions stationed in Chin State, and only two battalions operated there: Light Infantry Battalion (LIB) number 89 stationed in Kalaymyo, Sagaing Division, and LIB 50 stationed in Kankaw, Magwe Division. Chin State now hosts 14 battalions, with an average of 400–500 soldiers each and 50 army camps (Human Rights Watch 2009b).

The army has systematically used restrictions on fundamental freedoms, forced labour for military construction, torture, arbitrary arrests, unlawful and prolonged detention and attacks on religious freedom in Chin State to subvert any form of resistance against it. There have been many instances of arbitrary arrests, torture and disappearances perpetrated by the army relying on vague and broadly defined laws from the British era. In one instance of torture documented by Human Rights Watch, a 16-year-old Chin boy arrested for having links with the Chin National Army (can)—the Chin rebel group—described his encounter with the Tatmadaw thus:

(The army) beat me with a stick and they used the butt of their guns. They hit me in my mouth and broke my front teeth. They split my head open, and I was bleeding badly... They also shocked me with electricity....They would turn the electricity on and when I couldn't control my body any longer, they switched the battery off. They kept doing this for several hours.... They told me they would only stop beating us until we told them information about the CNA. We kept telling them we didn't know anything. (Human Rights Watch 2009b)

Although Chin had been travelling to Mizoram in India since the 1970s, fulfilling the demand of cheap labour because of the shared ethnic ties between the Mizo and Chin tribes, a surge of refugee flow began after 1988, and there are around 75,000–100,000 Chin in Mizoram (Levesque and Rahman 2008). A drive to remove foreigners, undertaken jointly by the Young Mizo Association (YMA) and the Mizoram Police after an incident of rape was committed by a Chin national on a Mizo girl and the perceived strain put by Chin on the state's economic resources, has led to several cases of eviction and repatriation. Although this is illegal according to international law, Mizoram state defines Chin refugees as 'economic migrants' and not as 'political refugees' (Chakma 2003). The Chin have sought refuge in various parts of the national capital too, with some registering with UNHCR to avail of their rights but most living in deplorable conditions, lacking an official refugee status.

THE ROHINGYA: THE MOST PERSECUTED STATELESS PEOPLE IN THE WORLD

The Rohingya are one of the most persecuted stateless people in the world. They were not accepted in any and faced expulsion from every country in their neighbourhood (ABC News 2015). India has been an exception, although only a small minority of Rohingya takes refuge in India (Khani 2015). To some, the Rohingya are a group of people who originally belonged to Bengal and subsequently migrated to Burma during colonial times, whereas another viewpoint states that they are those who hail from Arakan in Myanmar (Velath and Chopra 2015). The fact is that regardless of their origins, the Rohingya have been staying in Rakhine and some areas of Arakan for generations. Yet, they were never recognized as legitimate citizens of Burma.

Operation Dragon King launched by the Burmese army in 1977 to identify and deport foreigners resulted in widespread killings, rapes, destruction of mosques and religious persecution of 'illegal immigrants', culminating in the fleeing of 200,000 Rohingya from Arakan State to Bangladesh. Religious persecution of communal violence against and forced migration of the Rohingya continue to this day. Their shops, houses and fields have been burnt down, and countless human rights violations have been perpetrated against them with the tacit consent of state machinery and the ultra-orthodox clergy. 'In Burma, our parents were killed in front of our eyes, mothers were raped in front of their children, we were beaten like stray dogs', Mohammad Salim Ullah told Radio Free Asia in an interview conducted in a slum in northeast Delhi in May 2015 (Radio Free Asia 2015). According to UNHCR, in New Delhi alone there are about 17,000 Rohingya refugees. The registered asylum seekers from Myanmar are spread across the states of Andhra Pradesh, Delhi, Jammu, Haryana, Rajasthan and Uttar Pradesh. The conditions of most of these refugees remain deplorable, with no access to electricity, clean water and toilets. Most of them work as ragpickers or daily wage labourers, earning a meagre amount for self-sustenance (The Mahanirban Calcutta Research Group 2015).

SOCIO-ECONOMIC CONDITIONS: THE QUEST FOR RIGHTS

This segment deals with the socio-economic conditions of those communities that escaped to India and their socio-political mobilization to claim their human rights. It also highlights the consequences or effects of such quest.

THE BURMESE INDIANS AND THEIR SUCCESSFUL ASSIMILATION IN THE HOMELAND, INDIA

The literature on urban social movements shows that urban resettled citizens and migrants (and, therefore, returnees) are in a better position to organize and fight for 'rights to inhabit' a city (Castells 1983).

The migrants/refugees/returnees who settle in urban areas face a wide range of problems, such as housing, landownership and forcible eviction. Acquiring citizenship becomes essential to secure ownership rights and decent conditions for living. Realizing the need, returnee Indians have successfully organized movements on a variety of issues of concern to the repatriates, ranging from recognition of ownership of illegally occupied land parcels, the right to set up marketplaces or 'Burma Bazaars' and sell goods imported from Burma (and subsequently from elsewhere in Southeast Asia) and better 'rehabilitation programmes'. The returnees demanded governments to allot *pattas* (land entitlement) to all the Indian returnees from Burma under the Urban Land Ceiling Act (India) of 1968. However, owing to widespread corruption, lethargy and red tape, it took a long time for the returnees to secure land *pattas*. Some Indian returnees, such as the Tamil repatriates residing in B. V. Colony and the adjacent peri-urban settlements in the Vyasarpadi area (Shastri Nagar, Binny Garden and Indira Nagar) in Chennai, Tamil Nadu, got *pattas* only in 2008, after 43 years of endless petitions, meetings and negotiations led by the Burma Tamilar Munitra Sayalagam. Even in 2012, 10 per cent of the returnee families settled in B. V. Colony had not received the official *patta* (Egreteau 2013).

Manuel Castells, in his seminal work on urban social movements, observed that poor households and new migrants commonly push the boundaries of public restrictions in the urban spaces they have recently moved to, challenge cities' principal goals and set up informal—and therefore illegal—trading spots to earn a living. This is a political act, as the urban dwellers, knowing the illegality of their actions, organize themselves to counter the urban administrations attempting to regulate. The Tamil returnees from Burma formed an association named Burma Tamizhar Marumalarchi Sangam in Chennai to cater to the common interests and welfare of Burmese-Tamil shopkeepers, traders and merchants at the 'Burma Bazaar' near the port. The bazaar is considered as the largest open bazaar selling imported and smuggled foreign commodities. The association became necessary also to resist harassment from the police and the customs department, and to negotiate with the local urban administration the right to sell freely the commodities imported from Burma—and, later on, from elsewhere in Asia. Similarly, the Tamil merchants in Vellore formed the Vellore Burma

Bazaar Maruvazhvu Merchants Association. The Burma Repatriates Welfare Society (BRWS) in Ranhola suburb of Delhi has successfully negotiated its presence and acquired the right to agricultural lands. The 'Burma colony' at Ranhola recently became regularized in 2011 due to its persistent efforts (Egreteau 2013).

THE CHIN: THE 'OUTSIDERS' WHO STRUGGLE TO EKE OUT AN EXISTENCE

As the refugees in South Asia are encamped in close-knit groups where it is possible for them to exchange information, their mobilization becomes easier. A perception of collective suffering, combined with desperate living conditions faced by all, allows for the coordination required to make collective action possible. The support from agnate ethnic groups and international organizations also help the refugees. Apart from directly helping them or pressurizing the government to take action, international organizations give publicity and provide legitimacy to the refugees' cause. However, these roles have not been fulfilled by UNHCR in case of the Chin refugees for different reasons. The lack of external institutional support led to severe hardships for the Chin in eking out an existence. There have been reports of rape, exploitation, deplorable living conditions, diseases, such as tuberculosis (TB) and Hepatitis B, and malnutrition among children of Chin refugees living in India's capital city (Shepherd-Smith 2012). In Mizoram too—where Chin have stayed for longer and share ethnic ties with the locals—the Chin are labelled as 'foreigners' and have been targeted by local youth groups and the police.

THE ROHINGYA: LEAST MOBILIZED AND MOST VULNERABLE

The Rohingya are probably the least mobilized refugee group present in India, as their plight has only recently received international attention. Their entry into India from Burma is a recent phenomenon. They have been living in different cities. A majority of them do not possess a refugee status card (Velath and Chopra 2015). Some Rohingya youth work as daily labourers, but most depend on help provided by non-governmental organizations (NGOs). Health hazards and lack of basic education continue to affect them, as they cannot get access to hospitals and schools in the city. In some cities, like Hyderabad, different community organizations and NGOs are providing food, healthcare and other basic facilities to the Rohingya refugees who have taken shelter at the refugee camps at Balapur, Barkhas and Shaheen Nagar areas. Community organizations, such as Confederation of Voluntary Associations (COVA) and Civil Liberties Monitoring Committee, which help mobilize funds from local people, support the travel of Rohingya refugees to and from Delhi for obtaining a refugee certificate from UNHCR. Salamah Trust, a charity organization in Hyderabad, has set up the Burmese Refugees Relief and Rehabilitation Committee to provide shelter for refugees. It is also making efforts to provide education and healthcare. There are some such NGOs in Delhi and other cities too helping Rohingya refugees. However, the government's apathy and antipathy to the Rohingya refugees is rousing anxiety and insecurity among the Rohingya refugees in India.

INCONSISTENCIES IN THE STATE RESPONSE: LOOKING FOR A BROADER PARADIGM

The legal–institutional framework existing in India with respect to refugees is insufficient in dealing with their problems, as there is hardly any clarity, consistency and conviction in India's refugee policy. It is largely incumbent upon political considerations, geostrategic priorities and security implications. India has not ratified and is not a signatory to the Refugee Convention of 1951 and the Protocol of 1967 for a variety of reasons. The narrow definition of the term 'refugee', Euro-centric bias, preference for a bilateral approach to refugee issues and bureaucratic insensitivity coupled with a lack of political will are cited as the causes for India's ambivalence to the refugee problem (Chin Refugee Committee 2011). However, nonrefoulement as a principle has been accepted by the Indian government to handle certain refugee issues, and it has gradually secured the status of customary law. The critics point out that on the refugee problem, India has shown genteel indifference at worst and steadfast sympathy at best. For reasons of expediency and in the name of protection of national security, India has sometimes also applied the colonial-era legislations such as the Foreigners Act of 1946 and the Passports Act of 1967 to identify and deport refugees.

India has applied different yardsticks while dealing with refugees. Tamil refugees from Sri Lanka and Tibetan refugees have received considerable support from the state in terms of land and shelter, education and livelihood support and basic healthcare. Geopolitical and ethnic equations seem to be the reasons for India showing sympathy to the refugees from Sri Lanka and Tibet. In the case of Afghan, Iranian and Burmese refugees, India did not provide material or political support but allowed UNHCR to register and assist them. When it comes to India's response to the re-migration and refugee flows from Burma, one can see a significant variation in the state response. The Indian state has been sympathetic to Burmese-Indian 'returnees' and made efforts to resettle and re-integrate them into the national mainstream. The government took initiatives to facilitate their travel from Burma, resettled them in different states and initiated affirmative action to provide them relief (Mehrotra and Basistha 2011). Initially, in the refugee camps, the government provided food, shelter, medical allowance and education support for the children of the Indian returnees. The localities that they had illegally occupied and the businesses that they had started for making a living were legalized. In the course of time, the Burmese-Indian returnees got integrated with the Indian mainstream.

In contrast, the Chin and Rohingya have faced government apathy at best and police crackdown and torture at worst. The Asian Centre for Human Rights has documented the police brutality inflicted upon the protesting Chin in New Delhi leading to illegal detention, arbitrary arrests, torture and beatings in custody, and even attempts at repatriation (Zahau and Fleming 2014). The Chin have become easy targets of rapes, sexual exploitation and forced labour due to their 'outsider' tag. The Indian government has done little with regard to assisting the Chin in any meaningful manner and has shown glaring apathy towards their grievances. UNHCR cites a lack of adequate funds to support them, and their budget is dropping by the year. Those who receive a subsistence allowance from UNHCR are few and far between, with a large section of the Chin refugee population working as daily wage labourers or as workers in the informal sector in Delhi's many rubber, textile and other industries. The situation of the Chin in Mizoram is no different, and the resentful local police and youth organizations like the YMA have made life really hard for the Chin. Attempts were made to evict them from Mizoram by dehumanizing their sufferings, by calling them 'economic migrants' and by invoking the Foreigners Act, 1946.

As far as the Rohingya are concerned, the government has been indifferent to their plight. There is no official recognition of their existence. They face political insinuations, as if they are Inter-Services Intelligence (ISI) agents posing serious national security threat to India and hence there is a need to deport them. Apart from UNHCR, which assists them with a small allowance and provides refugee identity cards for stopping their repatriation, some community-based organizations and other NGOs are coming forward to help the Rohingya refugees. Provision of long-term visas and work permits for registered refugees is still not in place. In the state of West Bengal, more than 1,000 Rohingya refugees were prosecuted under the Foreigners Act, 1946, and sent to correctional homes. These people would be shifted to refugee camps only when they get refugee status by UNHCR.

CONCLUSION

Though India is not a party to the 1951 Convention Relating to the Status of Refugees, 1954 Convention relating to the Status of Stateless Persons and 1961 Convention on the Reduction of Statelessness, it is a party to several other important human rights instruments, such as the International Covenant on Civil and Political Rights (ICCPR), International Covenant on Economic, Social and Cultural Rights (ICESCR), Convention on the Elimination of All Forms of Discrimination Against Women (CEDAW) and Convention on the Rights of the Child (CRC). These human rights instruments make it obligatory for the Indian state to treat the aliens (refugees) in humane ways (Vijayakumar 2001). On the eve of its independence, India resettled millions of people from East and West Pakistan who came to the country for shelter. The treatment of Sri Lankan Tamils and Tibetans is another standout achievement of India's refugee experience.

However, India has been indifferent to certain categories of migrants/
refugees. The Chin and Rohingya have not had the same welcome that
Burmese Indians experienced during the 1960s–1970s. For the Indian
returnees, coming to India was more of a homecoming and easier
because of ethnic similarities with the mainstream Indian population.
In contrast, ethnic animosity and security concerns made India follow
a different approach in handling the Chin and Rohingya migrants/
refugees. It is time India recognizes the rights of the illegal migrants
and refugees, comes out with a national refugee policy and creates a
legal–institutional framework compatible with international refugee
conventions and protocols.

REFERENCES

ABC News. 2015. 'Myanmar to Deport Rohingya Migrants as United Nations
  Chief Urges Further Rescues'. Sydney, 24 May. http://www.abc.net.au/
  news/2015-05-23/myanmar-to-deport-migrants-as-un-chief-urges-further-
  rescues/6492618 (accessed 23 April 2016).

Bhaumik, S. 2003. 'The Returnees and the Refugees: Migration from Burma'. In
  Refugees and the State: Practices of Asylum and Care in India, 1947–2000, edited
  by R. Samaddar, 182–210. New Delhi: SAGE Publications.

Bobichand, R. 2015. 'Role of Civil Society and Media in India-Myanmar
  Relations: A Perspective from Manipur'. In Conference Report of the India-
  Myanmar Relations: Looking from the Border, 28–29 September. New
  Delhi: Heinrich Böll Stiftung. https://in.boell.org/sites/default/files/india-
  myanmar_relations_looking_from_the_border.pdf (accessed 23 April 2016).

Burma Human Rights Yearbook. 2006. 'Chapter 7: Rights of Women'. Mae Sot,
  Tak: Human Rights Documentation Unit of the NCGUB.' https://www.
  burmalibrary.org/sites/burmalibrary.org/files/obl/docs4/HRDU2006-CD/
  women.html (accessed 23 April 2016).

Castells, M. 1983. *The City and the Grassroots: A Cross-cultural Theory of Urban Social
  Movements.* Berkeley, CA: University of California Press.

Chakma, S. 2003. The status of children in India: An alternative report to the
  United Nations Committee on the rights of the child on India's first periodic
  report (CRC/C/93/ADD.5). Asian Centre for Human Rights, pp. 71–73.
  https://resourcecentre.savethechildren.net/node/2175/pdf/2175.pdf (accessed
  23 April 2016).

Chin Refugee Committee. 2011. 'Lives of Chin Refugees in Delhi: Case Studies'.
  New Delhi: Chin Refugee Committee. http://burmacampaign.org.uk/
  images/uploads/Case_Study_with_cover_for_email.pdf (accessed 23 April
  2016).

Dummett, M. 2010. 'Bangladesh Accused of 'Crackdown' on Rohingya Refugees.' *BBC*, London, 18 February. http://news.bbc.co.uk/2/hi/8521280.stm (accessed 23 April 2016).

Egreteau, R. 2013. 'India's Vanishing "Burma Colonies": Repatriation, Urban Citizenship, and (De)Mobilization of Indian Returnees from Burma (Myanmar) Since the 1960s'. *Moussons*, 11–34. http://moussons.revues.org/2312 (accessed 23 April 2016).

Ghoshal, D. 2013. 'Refugees Eke Out a Hard Life in India'. *The Atlantic*, New Delhi, 1 October. https://www.theatlantic.com/international/archive/2013/10/refugees-eke-out-a-hard-life-in-india/280132/ (accessed 23 April 2016).

Human Rights Watch. 2009a. 'Burma/India: End Abuses in Chin State India Should Offer Chin Refugees Protection.' Human Rights Watch, 28 January. https://www.hrw.org/news/2009/01/28/burma/india-end-abuses-chin-state (accessed 23 April 2006).

Human Rights Watch. 2009b. 'We Are Like Forgotten People: The Chin People of Burma, Unsafe in Burma, Unprotected in India.' Human Rights Watch, 27 January. https://www.hrw.org/report/2009/01/27/we-are-forgotten-people/chin-people-burma-unsafe-burma-unprotected-india (accessed 23 April 2016).

International Crisis Group. 2014. 'Myanmar's Military: Back to the Barracks?' International Crisis Group, 22 April. https://www.crisisgroup.org/~/media/Files/asia/south-east-asia/burma-myanmar/b143-myanmar-s-military-back-to-the-barracks.pdf (accessed 23 April 2016).

Khani, S. 2015. 'Islamic Countries Shut Door But India Welcomed Us: Rohingyas'. *Times of India*, New Delhi, 23 July. http://timesofindia.indiatimes.com/city/jaipur/Islamic-countries-shut-door-but-India-welcomed-us-Rohingyas/articleshow/48181384.cms (accessed 23 April 2016).

Levesque, J., and M. Z. Rahman. 2008. 'Tension in the Rolling Hills: Burmese Population and Border Trade in Mizoram'. IPCS Research Papers. http://www.ipcs.org/pdf_file/issue/1636771605IPCS-ResearchPaper14.pdf (accessed 23 April 2016).

Mehrotra, M., and N. Basistha. 2011. 'Collective Memories of Repatriates from Burma: A Case Study of West Bengal'. *Refugee Watch* 37:95–102.

Nivedita, R. 2015. 'Burmese Chin Refugees Protest at UNHCR'. *The Statesman*, New Delhi, 20 June. http://www.thestatesman.com/news/delhi/burmese-chin-refugees-protest-at-unhcr/70404.html (accessed 23 April, 2016).

Radio Free Asia. 2015. 'Delhi Slum Better than Life in Myanmar, Rohingyas Say'. New Delhi, 19 May. http://www.rfa.org/english/news/myanmar/india-rohingya-05192015101001.html (accessed 23 April 2016).

Shepherd-Smith, A. 2012. 'Bleak Prospects for Chin Refugees in India'. *IRIN News*, New Delhi, 21 June. http://www.irinnews.org/report/95699/myan-mar-bleak-prospects-chin-refugees-india (accessed 23 April 2016).

The Mahanirban Calcutta Research Group. 2015. 'Module C—Rohingya Refugees in India'. http://www.mcrg.ac.in/WC_2015/Module_C.pdf (accessed 23 April 2016).

Velath, P. M., and K. Chopra. 2015. 'The Stateless People—Rohingyas in Hyderabad, India'. The Mahanirban Calcutta Research Group. http://www.mcrg.ac.in/Rohingyas/Draft_Papers/Priyanca.pdf (accessed 23 April 2016).

Vijayakumar, V. 2001. 'A Critical Analysis of Refugee Protection in South Asia'. *Refugee* 19(2), January. http://refuge.journals.yorku.ca/index.php/refuge/article/view/22073/20741 (accessed 23 April 2016).

Zahau, R. and Fleming, R. 2014. 'A constant state of fear: Chin refugee women and children in New Delhi'. Chin Human Rights Organization (26 March). https://www.chinhumanrights.org/a-constant-state-of-fear-chin-refugee-women-and-children-in-new-delhi/ (accessed 23 April 2016).

Chapter 15

Drug Trafficking in and through Myanmar and Manipur

H. Srikanth and T. T. Haokip

INTRODUCTION

In 2003, the governments of Myanmar and India agreed in principle to undertake a joint survey to fence the Indo-Myanmar border. However, it was only in recent years, compelled by the common need to control ethnic insurgencies and check drug trafficking, that the two governments renewed their efforts to fence the international border. A Joint Boundary Working Group (JBWG) between India and Myanmar was constituted to examine/discuss all boundary-related issues comprehensively. According to the official reports, out of the 1,643-km-long border, demarcation of 1,472 km has already been undertaken. Only in two non-demarcated portions along the Indo-Myanmar border—one in Lohit sub-sector of Arunachal Pradesh, 136 km long, and the other in the Kabaw Valley in Manipur, 35 km long—is the work yet to be completed.[1] Government of India has initiated measures to construct 10-km-long border fencing in Manipur between border pillars 79 and 81 and approved ₹30.96 crore for the same. The Ministry of Home Affairs (MHA) claims to have completed the Detailed Project Report (DPR) and obtained clearances from the Ministry of Environment, Forest and Climate Change. The government has released the funds allocated, over ₹5 crore, for Compensatory Afforestation to the government of Manipur. It has released ₹16.38 crore to the Border Roads Organisation (BRO) for the fencing work.[2]

However, the initiatives of the governments to fence the border have met with resistance from the Kuki and Naga communities that reside in the hills along the Indo-Myanmar border (Das 2018; Oudot and Baudey 2018). In Manipur, the Kuki inhabitants of Govajang village near Moreh opposed the move, as they found that the proposed fencing project would bisect the village and the agricultural fields. The village chief approached the High Court of Manipur for compensation and stopped the government from constructing the fencing (Handique 2015). In Haolenphai, another nearby village, the villagers rose in protest when Myanmar Army personnel entered and destroyed the property, claiming the territory as part of Myanmar. The village chief complained that the villagers owing their allegiance to India had to suffer harassment, as Indian and Burmese authorities did not arrive at a consensus about the exact location of the original border pillar (Banerjee 2017). Reacting to these instances, the Kuki Inpi, the highest body of the Kuki people in Manipur, came out against the move to fence the border (Kuki Inpi 2013). Different Kuki civil and militant organizations have also demanded that the government reconsider its decision. Government of India is however determined to go ahead with the project. Against the background of these opposing stands or perceptions, this chapter attempts to highlight the problem of drug trafficking in and through Manipur and Myanmar and examine its implications for the security of the people in the borderlands of Manipur.

HISTORY OF THE INDO-MYANMAR BORDER

Prior to the Burmese colonization of Burma and the hills of Manipur, there were no restrictions on the movement of the indigenous communities inhabiting the borderlands of Manipur and Burma. The British stepped into the region on the pretext of protecting the Ahom and Manipuri rulers from the Burmese attack. After the defeat of Burma in the 1824–1826 war, the British annexed parts of Lower Burma and also gained control over Assam, Manipur and the Jaintia Hills. In

1834, the British Captain R. B. Pemberton drew the borderlines on the map between the Manipur Kingdom and Burma. After the Second Anglo–Burmese War (1852), the British took over Pegu province. Col James Johnston, the head of the Boundary Commission, surveyed the borders and erected border pillars in 1882 to demarcate the boundaries between the Burmese kingdom and British India. However, these lines on the map and the border pillars lost practical utility once the whole of Burma was annexed after the Third Anglo–Burmese War in 1885. Burma remained a province under British India for about five decades. When Burma was separated from British India in 1937, the Pemberton–Johnston line again became the border between British India and British Burma (Phanjoubam 2015; Pillalamarri 2017). Even at this point, the border did not affect the indigenous tribal people, as there was no fencing and the British did not try to impose any restrictions on the free movement of the people across the border during the colonial period.

After independence, in 1967, both India and Burma officially ratified the line as the international boundary (Ministry of External Affairs, GOI 1967). Although the border communities were never consulted in making the decisions about the borders, the national governments did acknowledge the presence of the border communities and agreed on a Free Movement Regime (FMR) enabling access to the indigenous border communities into the neighbouring state up to 16 km. As a result of the operation of the FMR, the border communities did not face any serious problem in maintaining relations with their kith and kin on either side of the Indo–Myanmar border. The Kuki and Naga communities in the state of Manipur continued to have familiar, social, cultural and economic interactions with their brethren in Myanmar (Haokip and Srikanth 2019). In the absence of fencing, the border pillars did not have any adverse effects on the daily lives of the communities, even though some border pillars were located in the middle of the villages and agricultural lands. However, the sense of security and freedom of movement that the indigenous communities have been enjoying since the colonial times appear to be under threat because of the changing security perceptions of the national governments of India and Myanmar wanting to fence the borders in the name of national security. Controlling the activities of insurgent groups and preventing the trafficking of drugs, arms and people across the border are cited as important factors necessitating the construction of border fencing.

MYANMAR: EPICENTRE OF DRUG TRAFFICKING

Myanmar is part of the infamous Golden Triangle, known for drug production and trafficking. Myanmar is the second-largest producer of opium after Afghanistan. It produces 25 per cent of the world's opium. Annual production of opium in 2012 was 690 tons, valuing about US$359 million (UNODC 2012). Poppy required for opium production is cultivated in Kachin and Shaan states in Myanmar, which are experiencing powerful ethnic insurgencies. In recent years, the Wa ethnic group has emerged as the largest producer of amphetamine-type stimulants (ATS). The United Wa State Army (UWSA) is the largest producer of methamphetamine (Lintner 2019). Drug barons have connections in different countries and even control the banking, airlines, hotel and infrastructure industries in Myanmar. Heroin and ATS produced in Myanmar are trafficked to the rest of the world mostly through Thailand and China. The government of Myanmar has passed strict laws against the production and trafficking of drugs, but it is not able to implement the laws in the border areas that are under the control of different ethnic groups. Security and conflict management, being the more important concerns, Myanmar's government does not pay much attention to the problem of corruption (International Crisis Group 2019). It has enacted strict anti-drug laws, but implementing them effectively becomes difficult because of the continuing insurgencies and corruption among mid-level military officers and government officials.

MANIPUR CONNECTION

However, in recent decades, due to continuous pressure from the international agencies like the United Nations Office on Drugs and Crime (UNODC), Myanmar's government has taken extensive measures to stop poppy cultivation in the areas of its control (UNODC 2017). Several thousands of acres of poppy fields were burnt down. As a result, the production of poppy seeds, which is essential for the production of heroin, stagnated in Myanmar, creating a demand for the same in the country. The demand for poppy seeds in Myanmar is tempting the peasants and villagers in the hill areas of Manipur adjacent to the Indo-Myanmar border to indulge in poppy cultivation in inaccessible hill areas. Poppy plants grow with little investment. Poverty, unemployment, underdevelopment and lack of alternate avenues for livelihood compel the villagers to go in for poppy cultivation (Kipgen 2019). Smuggling of raw poppy into Myanmar becomes easy, as there is no fencing along the Indo-Myanmar borders. Apart from poppy seeds, the production of heroin requires certain chemicals and pharmaceutical inputs, like pseudoephedrine (PH), which are not adequately available in Myanmar. Manipur has become a conduit for supplying these ingredients to drug cartels in Myanmar. Through different routes, these chemicals and drugs bought from mainland India enter Myanmar through Manipur. The police located many laboratories producing heroin in Myanmar's bordering Manipur state. Heroin, WY tablets and other psychotropic drugs produced in Myanmar are trafficked through Manipur to different cities in India and from there to Europe and other parts of the globe. According to Narcotics and Affairs of Border (NAB), heroin comes to Manipur through Myanmar before it is trafficked to other places. A kilogram of heroin, valued at around ₹4 lakh in Moreh, would cost up to about ₹90 lakh in Mumbai (Ghunawat 2017). Some known places in Manipur through which drug trafficking takes place are Moreh (Tengnoupal district), Behiang (Churachandpur district), New Somtal (Chandel district) and Tuisang (Ukhrul district) (Dhamen '90% of Drugs').

Traffickers use unique methods to carry drugs in and out of Manipur and Myanmar. They build specially designed vehicles for hiding the drugs and escaping the scrutiny of the security personnel. Apart from vehicle drivers and professional couriers, other local folks, such as unemployed youth, petty shopkeepers, village women, labourers and even children, are used to carry drugs from one point to another. Trafficking is no doubt risky, but many villagers and local youth engage in it, as it lets them make a quick buck. It is an open secret that although most insurgent groups outwardly denounce the trafficking of drugs, many of their grassroots cadres directly or indirectly indulge in poppy cultivation and drug trafficking (Nepram 2002).

GROWING DRUG ADDICTION IN MANIPUR

One of the side-effects of drug trafficking in Manipur is the growing drug addiction among the local people. Easy availability of drugs attracts many men and women to try drugs. There is a gradual shift from the oral consumption of alcohol and *gutka* to the consumption of heroin, WY tablets and other psychotropic drugs. Apart from oral drugs, addicts have started trying injected drugs. In Manipur, there are an estimated 45,000–50,000 drug addicts. School children, the youth and women—both rural and urban—are the victims (UNODC 2015). Studies have also revealed that 12 per cent of the drug addicts are below the age of 15, while 31 per cent are aged 16–25 and 56 per cent are aged 25–35.[3] Most of the drug addicts who use injected drugs are also victims of the human immunodeficiency virus (HIV). The number of HIV cases in Manipur is alarming. The increasing number of drug addicts and HIV cases led to the sprouting of several de-addiction centres in Manipur (Sehgal 1991, 26–28). The capital city of Imphal alone has around 20 de-addiction centres. The other two urban centres which have de-addiction centres are Churachandpur and Ukhrul.

INSTITUTIONS FIGHTING DRUG TRAFFICKING

Realizing the gravity of the problem of drug trafficking, Government of India enacted the Narcotic Drugs and Psychotropic Substances Act, 1985 and the Prevention of Illicit Trafficking in Narcotic Drugs and Psychotropic Substances Act, 1988. To realize the objectives and goals of these acts, the central government created the Narcotics Control Bureau (NCB) in 1986, with branches all over India, including in border areas of Northeast India. The government of Manipur created NAB as part of the state police department, with similar objectives as NCB. Apart from NCB and NAB, the Assam Rifles guarding the Indo-Myanmar border monitors the movement of vehicles on the Moreh highway and other outlets. In recent years, the state police has burnt down hundreds of acres of poppy fields and in collaboration with Customs department and NCB, conducted raids to catch drug peddlers and drug products. Now and then, one comes across the news reports claiming the seizure of drugs, arms and gold by the Assam Rifles and the state police. Some police officials, officers of Assam Rifles, local politicians and businessmen abetting drug trafficking were also caught.[4] Some of the drugs seized include amphetamine, the cannabis plant, cocaine, ephedrine, ganja, hashish, heroin, ketamine, lysergic acid diethylamide (LSD), acetic anhydride, 3,4-methylenedioxy-N-hydroxy-N-methylamphetamine (MDMA), methamphetamine, methaqualone, morphine and opium (Mallapur 2016).

The security personnel admit that what they could seize is insignificant compared to the actual volume of the drugs trafficked. Usually, those caught by the security personnel are petty drug peddlers who do not know the masterminds employing them. The police officials admit that they could not capture kingpins involved in the drug business. Trafficking does not take place only through the Moreh and Behiang routes. The drug traffickers make use of multiple path tracks along the Indo–Myanmar border, where there are hardly any security check posts. The drug traffickers move in and out of the open Indo–Myanmar border, taking multiple path tracks along the border where security checks are difficult. As the Assam Rifles is seen as a counter-insurgency agency and its branches are stationed far away from the border areas, there has been a demand in Manipur for delegating the responsibility of securing the borders to the more competent agency. Of late, the government is toying with the idea of creating an Indo–Myanmar Border Force (IMBF) with 29 battalions to police the borders (Gupta 2018). As part of the Look East/Act East policy, the central government is building an integrated check post at Moreh to facilitate and monitor all exports to and imports from Myanmar.

CONCLUSION

The insurgent activities along the Indo–Myanmar border were indeed one of the reasons that compelled the governments of India and Manipur to initiate steps for Border fencing. However, the series of peace treaties that India entered into with various ethnic insurgent groups have restored peace to some extent. There are still some Naga and Meitei militant groups that refuse to negotiate with the government. However, after the conclusion of the ceasefire agreements with Kuki militant groups, the borderlands of Manipur–Myanmar have become relatively peaceful. The valley-based Meitei militant groups are no doubt still active in the state, but their challenge has been considerably neutralized in recent years. Thus, compared to ethnic militancy, trafficking of gold and drugs has emerged as a more important problem for human security in Manipur. The governments have been emphasizing the threat of drug trafficking and drug addiction as the rationale for fencing the international border that remained open for over one century.

However, the change in the governments' policy of fencing the border is creating anxiety in the minds of the indigenous tribes inhabiting the Borderland. Where the boundary line passes through the middle of villages, as with Govajang village, the people's opposition is but natural. Similarly, in Haolenphai village, there are differences between Indian and Burmese authorities over the exact location of border pillars, and they accuse each other of moving the original border pillars. Barring some such controversies here and there, the Manipur–Myanmar border is comparatively less contentious and is more or less settled. Fencing of the borders may help to some extent in controlling the trafficking of drugs, arms and human beings. However, the border fencing alone cannot solve the problem. Inaccessibility, underdevelopment and lack of viable sources of livelihood in the hill areas facilitate drug trafficking in the border areas. The government should, therefore, make efforts to create employment opportunities and alternative sources of livelihood for the villagers bordering Myanmar. Simultaneously, with the help of non-governmental organizations and local traditional institutions, the governments should create awareness about the ill effects of drug addiction and rehabilitate drug addicts. More than anything else, before going ahead with the border fencing, the government should allay the fears and apprehensions of the indigenous communities living in the border areas. It is necessary to ensure that the border fencing would not adversely affect the free movement and everyday interactions of the indigenous communities living on either side of the Indo-Myanmar border.

NOTES

1. Government of India, Ministry of Home Affairs, *Annual Report 2018–19*, p. 40.
2. Ministry of Home Affairs, Government of India, *Annual Report 2016–17*, p. 41; Standing Committee on Home Affairs 2017.
3. The media in Manipur continuously report on the drug abuse and its effects. See, Felix, ASK. 'Drug Abuse Scenario in Manipur'. https://drugfreene. wordpress.com/2008/11/11/drug-abuse-scenario-in-manipur/; 'Udta Punjab to Udta Manipur: Drug Addiction Among Manipuri Women on the Rise.' 2016, 18 November. http://theresponsibleindian.com/women/from-udta-punjab-to-udta-manipur-the-alarming-drug-problem-in-manipuri-women 'Udta Punjab to Udta Manipur: Drug Addiction Among Manipuri Women on the Rise.' 2016. http://theresponsibleindian.com/women/from-udta-punjab-to-udta-manipur-the-alarming-drug-problem-in-manipuri-women (accessed 18 November 2016).
4. The criminal nexus between the policymakers and drug peddlers is reported in the media. See Samom (2013)

REFERENCES

Banerjee, Rabi. 2017. 'On Peril's Edge'. *The Week*, 14 May. https://www. theweek.in/theweek/current/india-myanmar-border.html (accessed 8 February 2021).

Das, Bidhayak. 2018. 'Dissent in the Naga Hills as India-Myanmar Follies Linger'. *The Irrawady*, 7 May. https://www.irrawaddy.com/features/dissent-in-the-naga-hills-as-india-myanmar-border-follies-linger.html (accessed 26 December 2019).

Dhamen, Thingbaijam. 2009. '90% of Drugs in India Trafficked from Myanmar Through Manipur'. *E-Pao*, 1 March. http://e-pao.net/GP.asp?src=14..010309. mar09 (accessed 12 August 2018).

Ghunawat, Virendrasingh. 2017. 'Heroin, Cocaine, Marijuana, Synthetic Drugs, Put India's Narcotics Markets on a High'. *Indian Today*, 21 December. https:// www.indiatoday.in/india/story/synthetic-drugs-push-indias-narcotics-market-to-a-high-1113532-2017-12-21 (accessed 8 February 2021).

Gupta, Shishir. 2018. 'Home Ministry Looks to Form 29-battalion Indo-Myanmar Border Force'. *Hindustan Times*, 18 January. https://www.hindustantimes. com/india-news/home-ministry-looks-to-form-29-battalion-indo-myan-mar-border-force/story-iImWizVCRAtDYGIonj2OML.html (accessed 14 November 2020).

Handique, Maitreyee. 2015. 'Govijang, A Village Without a Country'. *The Quint*, 26 June 2015. https://www.thequint.com/news/world/beyond-nation-and-state-in-an-india-myanmar-border-village (accessed 12 December 2019).

Haokip, T. T., and H. Srikanth. 2019. 'Fencing the Indo-Myanmar Border: Problems and Perceptions of the Kukis in Moreh and Tamu'. Report Submitted to ASEAN Studies Centre, Shillong.

International Crisis Group. 2019. 'Fire and Ice: Conflict and Drugs in Myanmar's Shan State'. Report No. 299, 8 January. https://www.crisisgroup.org/asia/south-east-asia/myanmar/299-fire-and-ice-conflict-and-drugs-myanmars-shan-state (accessed 23 December 2019).

Kipgen, Ngamjahao. 'Why are Farmers in Manipur Cultivating Poppy?' *Economic and Political Weekly* 54(46), 23 November. https://www.epw.in/engage/article/why-are-farmers-manipur-cultivating-poppy (accessed 24 December 2019).

Kuki Inpi. 2013. 'Open Memorandum to the Prime Minister of India and the President of Myanmar'. 23 September.

Lintner, Bertil. 2019. 'The United Wa State Army and Burma's Peace Process'. United States Institute of Peace, No. 147, April, 10–12.

Mallapur, Chaitanya. 2016. 'India's Soaring Drug Problem: 455 Percent Rise in Seizures'. Scroll.in, 26 June. https://scroll.in/article/736924/indias-soaring-drug-problem-455-rise-in-seizures (accessed 14 November 2020).

Ministry of External Affairs, GOI. 1967. 'Boundary Agreement Between the Government of India and the Government of the Union of Burma'. 10 March. https://mea.gov.in/bilateral-documents.htm?dtl/5886/Agreement+on+International+Boundary+with+India (accessed 8 February 2021).

Nepram, Binalakshmi. 2002. *South Asia's Fractured Frontier: Armed Conflict, Narcotics and Small Arms Proliferation in India's North-East*. New Delhi: Mittal Publications.

Oudot, Carole, and Matthieu Baudey. 2018. 'A Wall Goes Up in Myanmar'. *Asia Times*, 31 January. http://www.atimes.com/article/wall-goes-myanmar/ (accessed 26 December 2019).

Phanjoubam, Pradip. 2015. 'India's War Against Itself'. *Economic and Political Weekly* 50(25), June. https://www.epw.in/journal/2015/24/reports-states-web-exclusives/indias-war-against-itself.html (8 February 2021).

Pillalamarri, Akhilesh. 2017. 'When Burma was Still Part of British India'. *The Diplomat*, 30 September 30. https://thediplomat.com/2017/10/when-burma-was-still-part-of-british-india/ (accessed September 12 2018).

Samom, Sobhapati. 2013. 'MLA's Son Arrested in Manipur Drugs Case'. *Assam Tribune*, 28 February.

Sehgal, P. N. 1991. 'Shocking Findings in Manipur'. *Health Millions* 17(4):26–28.

Standing Committee on Home Affairs. 2017. 'Border Security: Capacity Building and Institutions'. No. 230. Report Submitted to Rajya Sabha on 11 April 2017. New Delhi: Rajya Sabha Secretariat.

UNODC. 2006. 'Executive Summary: Drug Use in the North-eastern States of India'. https://www.unodc.org/pdf/india/drug_use/executive_summary.pdf (accessed 14 November 2020).

UNODC. 2012. 'South-East Asia Opium Survey 2012: Lao PDR, Myanmar'. https://www.unodc.org/documents/crop-monitoring/sea/SouthEastAsia_Report_2012_low.pdf (accessed 13 December 2019).

———. 2015. 'Women Who Use Drugs in Northeast India'. https://www.unodc.org/documents/southasia/publications/research-studies/FINAL_REPORT.pdf (accessed 14 November 2020).

———. 2017. 'Myanmar Opium Cultivation Declines Sharply, Except in Some Conflict Areas: UN Report'. 6 December. https://www.unodc.org/unodc/en/press/releases/2017/December/myanmar-opium-cultivation-declines-sharply--except-in-some-conflict-areas_-un-report.html (accessed 25 December 2019).

Chapter 16

Understanding the Kukis' Opposition to Fencing of the Indo-Myanmar Border

T. T. Haokip and H. Srikanth

INTRODUCTION

The India–Myanmar border measuring 1,643 km in length is unique in different ways. The international border touches four north-eastern states, namely Mizoram, Manipur, Nagaland and Arunachal Pradesh. Although the borders were delimited during the colonial period, other than erecting the pillars, the British did not try to prevent the movement of the indigenous tribes inhabiting the borderlands. Even after independence, for over four decades, the governments of India and Burma (Myanmar) did not feel any need for fencing the border. Unlike the heavily fenced and guarded India–Pakistan and Indo-Bangladeshi borders, India's international border with Myanmar remained open. The 1952 Agreement between the two countries brought into force a Free Movement Regime (FMR) that permitted the indigenous border communities to travel up to 16 km into the neighbouring country with no visa restrictions. The FMR enabled the indigenous border communities to maintain their relations with their kith and kin on the other side of the border. However, a few years back, citing security concerns, the governments of India and Myanmar have taken steps to fence the border. The border communities—the Nagas and the Kukis—opposed the decision of the governments and have demanded the governments not to pursue the fencing project. The present chapter, based on the study undertaken as part of the ASEAN Studies Centre, Shillong, throws light on the interactions and interdependence that take place among the Kukis of Moreh (India) and Tamu (Myanmar) and gives their perceptions on the fencing of the border.

THE KUKIS: A BORDER COMMUNITY

The term 'Kuki' refers to a conglomeration of several agnate tribes settled in contiguous territories straddling the northwest of Myanmar, the Chittagong Hill Tracts in Bangladesh and India's north-eastern states (Vaiphei 1986, 27). The Kukis living in Manipur have racial, cultural and linguistic similarities with the Chins in Myanmar and the Mizos in Mizoram, and together they are referred to as the Kuki–Chin–Mizo group (Carey and Tuck 1987, 24; Shakespear 1982 (1912)). The tribes identified as belonging to the Kuki ethnic stock include Aimol, Anal, Biete, Chiru, Chothe, Darlong, Gangte, Hallam, Hmar, Khelma, Hrangkhawl, Koireng, Kom, Lamkang, Lusei, Moyon, Monsang, Paite, Pawi, Ralte, Simte, Tarao, Thadou, Zou, etc. (Grierson 1990; Shaw 1997 (1929), 16). Unlike the Naga tribes who speak different languages, one can see considerable linguistic and cultural similarities among the Kuki–Chin–Mizo tribes, which makes it easier for them to understand and communicate with one another. Before colonization by the British, the Kuki–Chin–Mizo tribes were spread over the western frontier areas of the (Burmese) Ava Kingdom, the Chittagong Hill Tracts, hills of Tripura and the hill areas of Manipur (Grierson 1967). During the pre-colonial period, in search of food and cultivable lands, the Kukis and other tribes used to move within the territory from one place to another. The territory acted as a buffer between the then Burmese kingdom and the areas that later became part of British India. Kuki scholars and social activists refer to these contiguous territories inhabited by the Kuki–Chin–Mizo tribes as 'Independent Kuki Country'.[1]

The colonial writers projected the Kukis as tribes that migrated to the territories only during the colonial period (Johnstone 1971[1897]; McCulloch 1980). However, the Royal Chronicles of Manipur acknowledge the presence of the Kukis even during the pre-colonial times (Kamei 1991; Paratt 2005; Phukhan 1992). Some writers consider the Kukis as among the oldest inhabitants in Northeast India. They lived in their respective villages in proximity to the Nagas and other indigenous communities. The Kuki villages had their traditional political institutions. Traditional Kuki polity, based on chiefdom, functioned with a full complement of governing bodies, such as Semang (home minister), Pachong (defence and external affairs minister), Lhangsam (minister of public relations and broadcasting) Lawm Upa (minister of youth, economic and cultural affairs), Thiempu (priest) and Tollai Pao (law and order enforcement minister) (Haokip 1995). After 1891, the British let the native Meitei kings rule the valley area but delegated the administration of the hill areas inhabited by the Nagas and Kukis to the political agent (Ibochou Singh 1985, 144–149). The Christian missionaries that worked in collaboration with the colonial government built churches and established schools for the hill tribes. The spread of Christianity, the growth of modern education and exposure to modern political and legal institutions brought changes in the Kuki society (Haokip 2016). However, unlike the Nagas, the Kukis continued to oppose the British colonization.

ANTI-COLONIAL HISTORY OF THE KUKIS

Often referred to as savage headhunters, these fiercely independent Kuki tribes were never subordinate to any feudal kings that ruled the plains. Since 1761, the Kukis in the hill areas of Manipur, the Chin Hills and the Kabaw Valley waged many a struggle to stop the British efforts to colonize the Kuki territory. It took several decades for the British to annex the hill areas inhabited by the Kuki–Mizo tribes.[2] The British took control of Kangleipak (Manipur) in 1891 and appointed a political agent to oversee the affairs of the hill communities. In their colonial interest, the British chose not to interfere with the rule of the Kuki chiefs in the hill areas. They did not meddle with the landownership and traditional administration. Barring certain developmental initiatives, such as the construction of footpaths and seasonal roads, the Meitei king and the British did not take up any developmental activity in the Kuki areas (MPCC (I) 1985).

Unlike other hill tribes in the Northeast whose resistance ended by the end of the 19th century, the Kukis remained hard nuts for the British to crack. Their resistance continued even in the 20th century. Unlike the Nagas and the other tribes that took part in the First World War as labour corps and took pride in fighting the war on behalf of the British, the Kuki chiefs fiercely opposed the British efforts to recruit Kukis in Manipur and Burma. The British had to use paramilitary forces and spent ₹28 lakh to suppress the Great Kuki Rebellion, 1917–1919.[3] L. W. Shakespear described it as the 'largest series of military operations' in the eastern frontier of India (Shakespeare 1981[1929], 235–236). After the First World War, efforts were made to tame the resilient tribes through Christian missionaries. The growth of Christianity did not however stop the Kuki chiefs from lending support to the Subhas Chandra Bose-led Indian National Army (INA) that fought against the British Raj during the second World War. After the war, the British arrested and punished the Kuki chiefs in Manipur and Burma who had lent their support to the INA (Kipgen 2015, 232–238).[4]

THE KUKIS AND THE HISTORY OF THE INDO-MYANMAR BORDER

Before colonization, the Kukis could move from place to place and interact with their kith and kin in other villages. There were no boundaries demarcated between the Manipuri and Burmese kingdoms. The enmity and wars between Manipuri kings and Burmese kings had a minor effect on the lives of the Kukis and other indigenous tribes. After taking control of Manipur, the British demarcated the boundaries of Manipur and Burma for security reasons. In the year 1834, the British authorities drew an imaginary line on the map between Burma and Manipur, which came to be known as the Pemberton Line, but it was only in 1882, after the survey by James Johnston and his party, that the British erected pillars demarcating the boundaries of Manipur and Burma. However, these border pillars did not make any difference, as the British colonized the whole of Burma by 1886. Till 1935, Burma remained a part of British India. After the Act of 1935, the British created a separate Burma Division in 1937, formalizing the separation of British India and Burma. Ten years later, the partition of British India into India and Pakistan led to the further division of the

Kuki territories. It is a matter of great concern that all these political developments that took place without the knowledge and consent of the indigenous communities made the Kuki–Chin–Mizo community the subjects of three countries, namely India, Myanmar and Pakistan (later Bangladesh) (Haokip 1998, 259; Kuki Movement for Human Rights 2009, 1).

Sociopolitical effects of the partition in 1947, the problem of the refugees during the Bangladesh Liberation War in 1971 and illegal migration that continued thereafter compelled India to go in for fencing the India-Bangladesh Border. The securitization of the Indo-Bangladeshi border virtually put an end to positive interactions between the communities on either side of the border. In contrast, the borders between India and Burma (later Myanmar) have remained open to this day. The 1952 Border Agreement allows the indigenous tribes living along the Indo-Myanmar border to cross the border and travel up to 16 km inside the neighbouring country without any passport (Thakur 2014). Although the relations between India and Myanmar were not always cordial, neither of them thought of fencing the borders for decades. However, the recent decision by the governments of India and Myanmar to fence the 1,643-km-long Indo-Myanmar border seems to have created ripples among the indigenous border communities in Nagaland and Manipur.

The state of Manipur shares 398 km of the border with Myanmar. The border areas in the Manipur districts of Tengnoupal and Churachandpur, bordering Myanmar, are inhabited predominantly by Thadou-speaking Kukis. On the other side of the border, in Myanmar, one can see habitations of the Kukis in and around Tamu. To understand the Kukis' opposition to the proposed border fencing, it is necessary to study the nature of relations and interactions that take place between the Kukis of Moreh and the Kukis of Tamu.

INTERACTIONS AMONG THE KUKIS IN THE BORDERLAND

The Kuki people in Moreh and Tamu have had ethnic, religious, economic and sociocultural ties for centuries. Before colonization, no boundaries separated them and obstructed their day-to-day interactions.

The border between Manipur and Myanmar was demarcated only during the British rule, that too without taking into consideration the wishes of the Kukis and other indigenous tribal communities living in the region. The border pillars erected during the British era were later endorsed by the two sovereign countries—India and Burma. However, this did not have much impact on the Kukis living in and around Moreh and Tamu, as the two governments agreed to allow access to the indigenous border tribes to travel up to 16 km on the other side of the border. To date, the Indo-Myanmar border has not obstructed daily interactions among the Kukis in the borderlands. The multiple ways in which the Kukis of Moreh and Tamu depend on and interact with each other in their daily lives are worth understanding.

Familial Ties

The Kukis in the borderland are connected by familial relations. Many Kukis in Moreh and Tamu stated that they have close relatives on the other side of the border. Marriages take place between the Kukis in Moreh and those in Tamu. They visit each other regularly on different occasions—at times of birth, marriage and other family functions and during illness, death and other troubled times. The absence of border fencing enables them to visit relatives and friends with no difficulty, even during the night. Different Kuki clans also maintain their genealogical tree and often organize clan meetings across the India–Myanmar border. It is through such clan meetings that they come to know the settlements of their relatives, in both India and Myanmar.

Cultural Relations

The Kukis are a close-knit society bound by a common culture, tradition, customs, affinitive dialects and a common past. Certain rituals and customs are observed by Kukis on both sides of the border. Mention may be made of the tribal customary practices such as *nao khose* (a married couple who have a child having to inform the wife's parents), *naopui* (the child being taken to the mother's father's house for blessings), *sating* (the flesh on the spine being given to the head of the clan by juniors) and *thitwibuh* (a bereaved family being visited with a jar of tea), which are practised by all Kukis despite their conversion to Christianity (Lenthang 2013, 73–78; Tuboi 2016).

The narratives of the Kuki people of India or Myanmar transcend the political boundary, as they share in common several cultural practices, folk tales and myths of origin and migration. Legendary tales of Kuki heroes and heroines, such as of Galngam, Khupting and Ngambom, Pujil and Langchal, Benglam, Jonlhing and Nanglhun, have regaled generations of Kukis across the border. The folklore have been passed down through the oral tradition. Further, Sa-Ai (a celebration of a successful big game hunt of big animals), Chang-Ai (a celebration of bounteous rice harvest), Hun (an occasion of worship in ancient times) and Chon le Han (hosting of an occasion involving feasting and sports) were observed across the border before the introduction of Christianity.

The Kukis in Moreh and Tamu believe that they all belong to the same nationality and have the same origins, history and destiny. They all speak and/or understand the Thadou-Kuki language and follow the same culture and traditions. Although the Kukis in Myanmar converse in Burmese and Meiteilon in Tamu, they speak in the 'Thadou-Kuki' language/dialect with the Kukis in Moreh whenever they meet in markets and other places. The festivals they celebrate and the rituals they follow are similar. Many Kukis from Tamu join the Kut Festival organized by the Kukis in Manipur. They take part in dancing, singing, fashion parades, sports, etc. Similarly, the Kukis in Moreh attend the Water Festival held in Tamu. During Christmas, the Kuki youth from either side of the border visit each other for fun and merrymaking. Such activities help in promoting cultural integration and awareness concerning their origins and identity and showcase the rich heritage of the people divided by political boundaries (Haokip 2003). Kut, a post-harvest festival is one such occasion where different Kuki–Chin–Mizo tribes across the border participate.

Religious Interactions

The Kukis living in the borderland are mostly Christians. Different denominations of Christianity have their associations in Moreh and

Tamu. The growth of Christianity and the establishment of churches offer the Kuki people platforms for interaction and discussion and fellowship among fellow Christians across the borders. The Kuki Church denominations in Tamu depend on their counterparts in Manipur. The Kuki Church in Tamu relies on the Christian literature published by the Kuki scholars in Manipur. The Kukis living in India and Myanmar share the same Sunday school literature and hymn books translated and published mostly from Manipur. The Church leaders in Moreh visit Tamu and address religious gatherings. Different seminars, workshops, leadership training courses and revival programmes organized by the Church denominations are attended by believers from Moreh and Tamu. Such inter-regional religious gatherings give opportunities for the young and the old to mingle and interact with one another. Many persons from Tamu areas came to Manipur for studying theology and other religion-related matters.

Opportunities for English Education

Burmese is the medium of instruction in educational institutions in Myanmar. English is hardly taught in Burmese schools. Many Kuki parents in Namphalong and Tamu aspiring for better career prospects for their children within the country and abroad send their wards to English-medium schools in Moreh. Hundreds of children from Namphalong and Tamu get admission in the schools in Moreh. Their Burmese identity is no bar for getting admission in the schools in Moreh. The security personnel rarely create any problem for the movement of the students as long as they are in school uniform and have valid identity cards. The educational institutions in Moreh become the meeting ground for the children of Moreh and Tamu.

Some Kukis from Tamu have travelled beyond Moreh, to Imphal, Shillong, Dibrugarh and even Poona and Bengaluru, for pursuing higher studies. When they go to distant places in India for higher studies, they identify themselves as Indians. Their relatives and friends in Manipur vouch for their Indian identity and even help them in getting valid identity cards to enable them to pursue studies. The authorities in India and Myanmar are not unaware of such movements, but they overlook them, as the security personnel understand the familial connections between the Kukis on either side of the border and do not see any threat to national security from such people. Usually, once they complete the studies, they go back to Myanmar and do not entertain any aspiration to settle in India leaving their family and property in Myanmar. One Kuki from Tamu who studied in Biblical Seminary College in Bengaluru informed the research team that the only problem he faced in moving into and out of India was communicating to the Indian border security personnel at Border Gate No. 2, who questioned him in Hindi. Being a Kuki from Myanmar, he said, he does not know any language spoken in India, other than Kuki and broken English.

Medical Needs

Moreh has a civil hospital with basic infrastructure. There are some private practising doctors. However, most Kukis say that the hospitals in Moreh do not properly take care of the patients. Hence, many of them rush to the hospitals in Tamu, where they think the doctors are more professional and caring. However, in cases of serious illness, the Kukis take the patients to the hospitals in Imphal. Even patients from Namphalong and Tamu are taken to the hospitals in Imphal to treat serious ailments.

Economic Interdependence

One can also see economic interdependence between the Kukis of Moreh and Tamu. Several Kukis from India take up *jhum* cultivation on Myanmar's side of the border. In contrast, the Kukis of Myanmar seldom cultivate in the Indian side of the border, as their lands are more fertile. In the early 1960s, several Kuki families from the Indian side moved to the plain areas in Myanmar and settled there to cultivate lands. Initially, there was no opposition to the Indian Kukis coming and cultivating the barren lands in Myanmar. Earlier, the Burmese government engaged them in clearing the forests and even leased land for cultivation for a few years, so that the government could use the land later for plantation and other purposes. However, of late, citing militant activities, Myanmar's government has imposed restrictions on the migrant cultivators from India. Still, as there are very few cultivable lands in the hills, several Indian Kukis cross the border and work as tenants in the lands of the Kukis in Myanmar. The Kuki landowners in Namphalong and Valpabung villages in Myanmar give protection to the Indian Kukis working in their lands by projecting them as Kukis from Myanmar. Further, the Kukis in Moreh depend on the forest products from Myanmar's side such as leaves, bamboo shoots, bamboos and logs. One can see several sawmills in Kuki villages on the Indian side of the border, the logs for which come partly from the forests of Myanmar. The Kukis in Chavangphai village in Moreh admit that some families belonging to the village take their cattle for grazing in the lands on the other side of the border.

Trade and Business

Even during the pre-colonial period, barter trade had been taking place between the border communities. Burmese villagers have traditionally bought cloth, medicines and other consumable items from Indian border *haats* or local markets, whereas fish, agricultural produce and some herbs are brought from the other side. This practice continued unhindered even during the British period, and after Burma and India became independent. The Kukis in Moreh frequently go to Namphalong and Tamu for shopping. They buy different types of fancy items and essential goods from Namphalong and Tamu markets, both for personal use and for sale in Moreh. Some take them to Imphal and other urban areas in the Northeast for sale. The Kukis in Namphalong and Tamu depend on India for medicines, motor vehicles, chappals, potatoes, cycles, betel nuts, *atta* (wheat flour), electronic gadgets, building material, tin roofs, etc. One can see Indian bicycles and motorcycles in many houses in Namphalong and Valpabung. It is said that many young Kuki men in Moreh and Tamu are involved in illegal trade, drug trafficking and smuggling activities, which assure them a quick buck.

To sum up, there has been a close bond between the Kukis of Moreh and Tamu. The international border did not obstruct their day-to-day interactions, as they could legally go to the nearby villages with no problem. The absence of border fencing enabled the Kuki villagers to cross the border and meet their kith and kin on the other side of the border any time they wished. They were no doubt aware of the presence of the border pillars, but they did not expect that one day there would be a fence connecting the border pillars which would restrict their movement across the borders. Naturally, the governments' decision to fence the international border has created panic among the Kukis living on either side of the border.

THE KUKIS' PERCEPTIONS ABOUT BORDER FENCING

Having lived for centuries with no border restrictions, the decision of the governments of India and Myanmar to fence the border has come as a rude shock to the indigenous tribes living on either side of the Indo-Myanmar border. The decision has been opposed by both Naga and Kuki chiefs. The research team, which surveyed different villages in Moreh and Tamu, interacted with select villagers, Kuki chiefs, Kuki militants, youth leaders and prominent Kuki personalities in Moreh and Imphal to elicit their opinions on border fencing. Different representations and memoranda of different Kuki organizations made to the authorities are examined. Some opinions expressed by the individuals and associations are presented here to enable the readers to understand the pulse of the community on border fencing.

In an 'Open Memorandum' addressed to the prime minister and the president of India, the Kuki Inpi of Manipur, Nagaland and Assam pointed out that the decision of erecting a few kilometres of fencing in Moreh–Tamu would affect 38 villages—19 villages in India and 19 in Myanmar territory. The decision would affect the entire 1,643 km of the international border, hundreds of villages and thousands of indigenous tribal peoples (Kuki Inpi 2013). Supporting the contention of the Kuki Inpi, in a memorandum addressed to the president of India and the president of Myanmar, Dr T. Lunkim, Chairman of Kuki Organization for Human Rights, also argued:

The then government of India failed to understand that the 90% land of the buffer zone is jointly owned by indigenous people through hundreds/thousands of Tribal villages under their Haosa (chiefs), Lals, Kalims, Khullakpas, etc. All these frontier areas of over 40 million population are still owned by the indigenous people. The proposed boundary fencing being constructed in the middle of their land is an open declaration to grab (neo-colonialism) the ancestral lands of the indigenous peoples of the so-called "independent Hill Country (sic)".[5]

In an interview with the research team, Dr S. Chongoi, Secretary of Kuki Organization for Human Rights, explained:

We live on our soil. The entire area belongs to us. We have not migrated to this land after independence. We live in our land from time immemorial. Our people know nothing about the independence of India, Burma, or Bangladesh. The governments partitioned our territory, declaring one side as Myanmar and the other as India or Bangladesh. They divided us. They did not ask us for our view and opinion. It violates human rights. In the middle of our land, they make the partition. But our people live on both sides of the border. We do not know where the border lies. The Kukis are in India, Myanmar, and Bangladesh. We are oppressed in all countries.[6]

Apart from the Kuki organizations, the move to fence the Indo-Myanmar border was criticized by Lal Thanhawla, former chief minister of Mizoram. In a letter addressed to the president of India, Lal Thanhawla wrote:

At present, the fencing work is being carried out most arbitrarily. Unfortunately, many villages inhabited by tribes of the same ethnicity are being cut-off from one another. Worse still, it is learned that in a few villages, even members of the same family and clan are being cut off from their relatives by this arbitrary fencing. This is a reality which is not peculiar to a certain village in Moreh area alone but is rather, a threat relevant to all parts of our region wherein our indigenous people of common ethnicity especially of the Mizo-Kuki-Chin kinship dwell on both sides of the international border which itself was in the first place demarcated arbitrary by the British. We have experienced a similar problem of fencing in the western part of Mizoram along the Bangladesh border.[7]

The village elders with whom we interacted expressed the fear that border fencing would adversely affect the day-to-day interactions between Kuki communities on either side of the border. In an interview, the chief of Haolenphai village in Moreh said,

> Border fencing will adversely affect our day-to-day lives since we are socio-culturally and economically dependent on our neighboring Kuki villages of Myanmar. Further, it will hamper our free movement across the borders, especially in times of sickness, sorrows, and happiness. The completed stretches of the border-fencing have already hampered to a great extent to our day-to-day lives. Earlier, we were able to move across the borders freely through short cuts and traditional footpaths. But now, we have to go through the officially designated gates.[8]

In a focus group discussion that the research team organized, one village elder from Chavangphai village, Moreh, said:

> We are divided by international borders, but we are conscious of belonging to the same community (Kuki). We migrated from China and settled in the Chin Hills. Then we migrated to Manipur from the Teddim-Chin Hills area. Our heritage belongs to the Teddim-Chin Hills. Our customs and traditions evolved from there. So, we have close cultural and kinship ties with the Kukis of Myanmar. Till today, most of us have our relatives there in Myanmar. If the government fence the border, the Kukis will be adversely affected. Therefore, I am strongly against the border-fencing project by India and Myanmar.[9]

The villagers in Valpabung (Myanmar) also echoed similar fears. For example, in a focus group interaction that we had in Valpabung, one lady in the village expressed her problem:

> I run a piggery farm in my home. I used to cross the border freely and at any point to get foodstuffs, vegetables, and forest products for my pigs from the other side of the border. But now that a stretch of the border fencing has been completed, I had to cross the borders through the check gates. This has adversely affected my daily activity and piggery farm. It will be worst when the border fencing is entirely completed.[10]

During our stay in Moreh, we could interact with some representatives of Kuki militant organizations. Mr Peter Mate, Joint Secretary, Home Affairs, Kuki National Organization (KNO), countered the argument that fencing is in the interest of the border communities. He said:

> The Kukis living in India and the Kukis living in Myanmar don't need border fencing. Fencing creates hindrances to the Kukis who wish to meet their relatives in times of death, illness, etc.... The decision to fence the border is taken without the consent of the Kukis....Why is that Myanmar becoming conscious of borders? Myanmar government wants to divide the Kukis settled in the borderlands. They want to stop interaction between the Kukis living in Moreh and Myanmar and weaken the Kuki people's movement. KNO is against fencing, and it has sent a memorandum to the Government of India. The Kukis can protect themselves, there is no need for fencing....[11]

Among the Kuki people we interviewed, a few remember that at one time the Kuki militant organizations themselves sought border fencing to protect the community from valley-based militant organizations that took refuge on the other side of the border. One young man in Chavangphai village, for example, cautioned:

> We cannot just speak against the border-fencing. A serious debate needs to be undertaken on the issue. It is imminent that valley-based Meitei insurgent groups are planning to enter Suspensions of Operations (SoOs) Agreement with the Government of India. In such a situation, Meitei insurgents will have a free-hand to move across the borders. The border -fencing is likely to check their free -movement across the borders and it will help to a significant control in law and order problems in the border areas. Even though we are now against the border fencing, it is worth reminding that in the first place, initially, we (the Kukis) have demanded border-fencing to check the law and order problems created by the Valley-based Meitei insurgent groups in and around Moreh. We should not oppose the border-fencing now.[12]

In our interactions, we came across some young educated Kukis who were conscious of the debates taking place at the global level on the rights of indigenous peoples. One young student in Haolenphai village insisted on innovative solutions to the border problems. He asserted:

> As the Kukis are settled in the territory now claimed by both India and Myanmar, both governments must ensure that the demarcation of boundaries would not obstruct the free movement of the Kuki community from Myanmar to India and vice versa. It is the responsibility of the civilized governments to respect and honor the rights of the indigenous people. India and Myanmar should draw lessons from the Sami, Kurd, and Yunnan experiences and create mechanisms that reconcile the national interests with the interests of the indigenous peoples.[13]

CONCLUSION

The study shows that there have been considerable fears and apprehensions among the Kuki masses about the proposed fencing. The Kukis on both sides of the international border express the fear that the fencing would obstruct their free movement and come in the way of familial, religious, sociocultural and economic interactions. They apprehend that as the border gates are not located near the villages, they would have to travel long distances to reach villages on the other side of the border. As the gates are manned by security personnel and are likely to be closed during nights, the villagers will face difficulties and are likely to be harassed by the security guards manning the border gates. They fear that once fencing is completed, they would no more have the freedom to take to traditional pathways to reach out to their kith and kin on the other side of the border. They apprehend that fencing creates obstacles in reaching out to their relatives in times of death, illness or such other tragedies. Naturally, the Kukis in the border villages oppose fencing the international border.

The Kuki militants and Kuki intellectuals that we spoke to accuse the governments of India and Myanmar of pursuing what they call 'colonial policies' with regard to the indigenous peoples. They accuse the governments of deciding on fencing without consulting the indigenous peoples living in the borderlands. They strongly oppose fencing that would divide the communities, villages and homelands and make them suffer for no fault of theirs. Although the Kuki militants and the intellectuals acknowledge the problem of drug trafficking and the need for protecting people from drug abuse, they consider that fencing is no solution to the problem. They say that as long as officials remain corrupt, no fencing could stop the trafficking of drugs and ammunition across the border. When it was pointed out to them that at one time the Kuki militants themselves were in favour of fencing to check the activities of the Meitei militants in the region, the militant leaders argued that to have or not have a fence is to be decided by the people, not by the governments. The border people should be taken into confidence and their fears and apprehensions should be addressed before taking any initiative to fence the international border.

NOTES

1. In all their memoranda, the Kuki organizations such as Kuki Inpi, Kuki Organization for Human Rights and Kuki insurgent groups referred to the Kuki-inhabited areas as 'Independent Kuki Country'.
2. While the Manipur Valley inhabited by the Meiteis was conquered in 1891, the hills inhabited by the Kukis were brought under British India and British Burma after the Anglo-Kuki War of 1917–1919.
3. Khaikhotinthang Kipgen called the war as 'Thadou War of Independence'. Gangte (2013) called it 'Anglo-Kuki War'. Haokip (1998), in his article, described the Anglo-Kuki War as the first Kuki war of independence.
4. See, Memorandum of the Kuki Political Sufferers Association, Manipur submitted to the Prime Minister of India in December 1958.
5. Interview with Lunkim, Imphal, 29 January 2018.
6. Interview with S. Chongloi, Imphal, 30 January 2018.
7. Lal Thanhawla's letter to the President of India, 1 November 2013.
8. Interview with Chief of Haolenphai village, 28 January 2018.
9. Focus group discussion at Chavangphai village, 27 July 2017.
10. Focus group discussion at Valpabung, 26 July 2017.
11. Field interview, 27 July 2017.
12. Focus group interaction at Chavangphai village, 25 July 2017.
13. Focus group interaction at Chavangphai village, 25 July 2017.

REFERENCES

Carey, Bertram S., and H. N. Tuck. 1987 (1932). *The Chin Hills, Vol. 1—A History of the People, British Dealings with Them, Their Customs and Manners, and a Gazetteer of Their Country*. Delhi: Gian Publishing House.

Gangte, T. S. 2013. *Anglo-Kuki War Relationship from 1849 to 1937 and Other Essays*. New Delhi: Ruby Press & Co.

Grierson, G. A. 1967. *Tibeto-Burman Family: Vol. 3*. Delhi: Motilal Banarsidass.

———. 1990. *Linguistic Survey of India*. New Delhi: Low Price Publication.

Haokip, Nehkholen. 2003. Jubilee KUT Celebration at Moreh—2003. November 9. http://kukiforum.com/2003/11/jubilee-kut-celebration-at-moreh-2003-2/ (accessed 18 December 2018).

Haokip, P. S. 1998. *Zale'n-gam the Kuki Nation*. Zale'n-gam: Kuki National Organization.

Haokip, T. T. 1995. *Kuki Chiefship and Its Changing Dimensions* (unpublished thesis). Shillong: Department of Political Science, North Eastern Hill University.

Haokip, M. Thongkhosei. 2016. *Ecumenism among the Kukis of North East India*. Secunderabad: GS Media.

Ibochou Singh, K. 1985. *British Administration in Manipur, 1891–1947* (unpublished PhD thesis). Guwahati: Guwahati University.

Johnstone, James. 1971 (1897). *Manipur and the Naga Hills*. Delhi: Vivek Publishing House.

Kamei, Gangmumei. 1991. *History of Manipur*. New Delhi: National Publishing House.

Kipgen, Seikhohao. 2015. *Political and Economic History of the Kukis of Manipur*. Delhi: Akansha Publishing House.

Kuki Inpi. 2013. Open Memorandum to the Prime Minister of India and the President of Myanmar. September 23.

Kuki Movement for Human Rights. 2009. *The Plight of the Indigenous Kuki People: Unraveling the story of Deception, Suppression and Marginalization in the Tri Border Area of India, Myanmar & Bangladesh*. Imphal: Kuki Movement for Human Rights.

Lenthang, Khuplam Milui. 2013. *The Wonderful Geneological Tales of Manmasi (Kuki-Chin-Mizo)*. New Delhi: Maxford Boooks.

McCulloch, W. 1980. *An Account of the Valley of Manipore and of the Hill Tribes*. Delhi: Mittal Publications.

MPCC (I). 1985. *Freedom Fighters of Manipur*. Imphal: Freedom Fighters Cell.

Paratt, Saroj Naili (trans). 2005. *Cheitharon Kumpapa: The Court Chronicle of the kings of Manipur*. London: Routledge

Phukhan, J. N. 1992. 'The Late Home of Migration of the Mizos', Presented in International Seminar on Studies on *Minority Nationalities of Northeast India—The Mizos*, Aizawl.

Shakespear, J. 1982 (1912). *The Lushai-Kuki Clans*. Cultural Publishing House.

Shakespeare, L.W. 1981 (1929). *History of the Assam Rifles*. New Delhi: Spectrum Publications.

Shaw, William. 1997 (1929). *Notes on the Thadou Kukis*. New Delhi: Spectrum Publications.

Thakur, Sajay. 2014. 'Indo-Myanmar Border Linkages'. October 5. http://www.claws.in/1264/indo-myanmar-border-linkages-sanjay-thakur.html (accessed 9 November 2018).

Tuboi, Paochon. 2016. Socio-cultural life of the Kukis in Manipur (unpublished PhD thesis). Shillong: North-Eastern Hill University.

Vaiphei, Prim. 1986. 'Who We are/Who are We?' In *In Search of Identity*, edited by H Kamkhenthang. Imphal: Kuki-Chin Baptist Union.

Chapter 17

Borders That Divide
Naga Ethnoscape and Idea of Supranational Citizenship

Jelle J. P. Wouters

Pierre Bourdieu, the French sociologist, pointed out how social action can end up 'producing a difference when none existed' and that 'social magic can transform people by telling them that they are different' (1993, 160–161). From this follows that difference (e.g., sacred/profane, success/failure, high/low caste, gender roles or, for that matter, any us/them categorization) is always a social imposition, designed by a social world that is subsequently constituted, and operates, through these differences. This social dynamic is particularly pertinent in the delineation of political borders, which for that reason are among 'the most paradoxical of human creations' (Gellner 2013, 5). In its formation, after all, borders often join what is different and divide what is similar (Van Schendel 2005, 9). But no matter how arbitrary and capricious the place and pretext of their creation, once borders are laid down, patrolled and recognized in terms of dividing lines, they release 'social magic' in that the social groups they enclose may acquire—especially as time passes—derivative signification, identity and identification from these borders (Wouters 2016). In this process, borders can become a convincing basis for the construction of contrasting identities, even when they dissever what was earlier conceived of as connected and similar.

Such a sequence, of borders reworking identities, reveals itself for-cibly among the Garos, whose ancestral lands are presently cut across by the Indo-Bangladeshi border. While a social divide between the Garo Hills (now in Meghalaya, India) and the Garo plains (now in Mymensingh division, Bangladesh) existed prior to the partition in 1947, their definite enclosure into two different nation states sharpened this divide. Bal and Chumbugong (2014, 96) observe thus: 'the differ-ent strategies of the Indian and Pakistani/Bangladesh states in dealing with their "peripheral" populations may have impacted local processes of self-identification and self-assertion in significantly different ways.' In broad terms, the Garo continue to identify as a single 'imagined community', yet Bangladeshi Garos today 'clearly identify as Garos *in* and *of* Bangladesh' (Bal and Chumbugong 2014, 97, emphases in original). Indian Garos, in turn, 'have developed ideologies of ethno-nationalism that do include territorial claims; claims, however, that exclude Bangladeshi Garos' (Bal and Chumbugong 2014). What this evidences, in short, is that boundaries differently circumscribe territories in terms of government and governance, development and policies of recognition and rights, besides, of course, nationalistic politics, and that this can produce diverse political identification and signification within a putatively singular 'ethnoscape'.

This is, however, not the only possible dialectic between bor-ders and identities—of the latter, that is, conforming to the former. Regardless of how formative and transformative exercises of border production can be in terms of social and political imagination, and further regardless of the very real effects they have on the governance of the contemporary historical moment, everyday flows and social processes habitually transgress such borders, even when states try to prevent such crossings through declaring them illegal. The differences they seek to inscribe and institutionalize, for one thing, can, and regu-larly are, undermined through everyday social networks, as happens in the Bengal borderland (Van Schendel 2005), and always also carry the potential to be dissolved in a different situation and time. In the context of the Nagas, my focus here, and contrary to the case of the Garos, there persists a clear subaltern agency and resistance in the con-joint attempt by the Indian and Burmese states to enclose, and close off, notions of Naga identity, belonging, history and ancestral territory, and in long-standing exertions of traditional authority, trade, tribute and other social networks into two nation states.

Over 70 years of post-colonial time and the influence of Indian and Burmese nationalist regimes have undoubtedly worked to somewhat blend current political boundaries with political imagination among the Nagas on either side of the Indo-Myanmar border. However, among them, Indian and Burmese nationalist tropes and representations have never precluded other, and non-statist, forms of belonging, identification and territoriality which refuse to limit themselves to these national spaces. The implication for ethnographic and social theory for the Nagas is that, for one thing, their study within conventional national frames—what is called the fallacy of 'methodological nationalism' (Gellner 2012)—conjures a deeply problematic, ahistorical and overtly statist approach. Here, instead, the study of ethnicity and belonging must 'go transnational'—not, to be sure, because Nagas are diasporic, but because nascent political boundaries arbitrarily truncated their ancestral lands and territories and long-standing sociopolitical networks between two different countries.

The decades-old, now highly factionalized and much-written-about Naga movement for the right to self-determination persists in its rejection of the division of Naga lands and peoples. It insists (at least formally) on the territorial and political integration of all Naga territories that currently lie divided between India and Myanmar and, within India, are now spread across four states: Nagaland, Assam, Manipur and Arunachal Pradesh (Wouters 2018). Naga nationalists invoke the term 'Nagalim' (Naga land) to protest this fragmentation of a contiguous Naga-inhabited area. This demand for territorial and ethnic tribal integration is not without its internal complications, contestations and conspiracies, as I have sought to detail elsewhere (Wouters 2016). I will not revisit these arguments here, except, that is, for two broad comments. First, Nagas' now long standing territorial and politico-administrative divisions have led to a colloquial distinction being regularly made today—in terms of identity, belonging and loyalty and patriotism—between 'Burmese Nagas' and 'Indian Nagas', as well as between 'Nagaland Nagas', Manipur Nagas', 'Assam Nagas' and 'Arunachal Pradesh Nagas.' While this does not negate the simultaneous, and evocative and emotive, projection of Nagalim, it does suggest that existing political borders have made an entry into the interstices of the sociopolitical imagination of 'Naga-ness'.

Second, there is considerable debate and disagreement about who is, and who is not, a Naga, and about what is, and what is not, ancestral Naga territory. To illustrate: while Nuh (2006, 24–26) is able to list 68 Naga tribes, Vashum (2005, 10) only counts 'forty odd Naga tribes', the Naga historian Horam (1975, 27) talks about '32 known Naga tribes', while the National Socialist Council of Nagaland (NSCN-IM), on its website (now offline), states that there are 43 Naga tribes. A similar confusion abounds approximations of Naga territory. Nuh (2002, 12) speaks about 47,000 square miles of Naga territory, Ao (2002, 6) estimates nearly three times as much—120,000 square miles—while Chasie (2005, 15) calculates Naga territory as ranging between 20,000 and 30,000 square miles, though probably 'nearer to the latter figure, if no more.' Such contradicting claims, rather than the net result of sloppy research, are perhaps best understood as axiomatic of the conjectures, politics and indeterminacies of the Naga nation-in-the-making, a process of enclosing and moulding disparate villages, village clusters and tribes into a generic Naga identity (Banerjee and Athparia 2004, 79; Lintner 2011, 112; Wouters 2018, 38–80). Such complications, again, may not undermine the assertion of a shared Naga identity and territory across political borders—in the end, all modern identities are historically contingent, constructed and continually debated—but they do complicate Naga nationalistic rhetoric that reconstructs and promotes a Naga historical consciousness of a clearly defined and distinct people that traditionally flourished in an equally clearly delimited territory.

The remainder of this chapter is modest in its ambitions. What it offers are a few broad insights and remarks regarding Naga political and territorial predicaments. The first section adopts a historical lens to illustrate the prolonged absence of political borders among Nagas, and in the adjoining highlands. It emphasizes that the very existence of political borders, in their contemporary sense, is a novel manifestation in the region, and that so, consequently, are the problems and sufferings they impart. Next, I discuss how the imposition of hard and fixed political borders by colonial, and especially post-colonial, states resulted in the present-day mismatch between political and ethnic boundaries between Nagas. The third, and final, section offers a preliminary suggestion to overcome, at least in parts, the current territorial division of the Naga homeland between India and Myanmar through discussing the idea and possibility of Naga 'supranational citizenship'.

WHEN THERE WERE NO BORDERS

Once upon a time, not so long ago, there were no borders in highland Northeast India, and in the adjoining hills—at least not in the contemporary sense of precise political lines with checkpoints, customs offices and border security guards at both sides, often eye to eye, engaged in a permanent quest to protect territorial sovereignty (What, after all, is border patrolling if not a physical and ritual (re)enactment of the territorial encompassment of the nation by the state?) and to prevent peoples and goods from crossing 'illegally'. Illegality, here, is a state category, not necessarily a social one, as what is deemed illegal may not always be considered illicit by borderlanders themselves (Abraham and Van Schendel 2005).

In much of this upland region, and certainly among the Nagas, the state, however, is a fairly recent arrival. Even more recent are nationalist teachings that 'state', 'nation' and 'border' *should* inspire passionate patriotism in defence of a state's territorial sovereignty, a vague idea that here long met a reluctant audience. This is not to suggest the historical absence of chauvinistic territoriality locally; Naga villagers, for one, were particularly protective of their fields and forests (and continue to be so), and ever anxious about possible intrusions and encroachments by 'outsiders', but this was a sentiment vested in deep material and spiritual attachments and affection for ancestral land, not in the 'meaningless' patrolling and defence of an abstract territory.

'Infiltration bids', 'border patrols', 'boundary pillars', 'smuggling', 'cross-border terrorism', 'illegal immigration', 'visas' and 'customs'—these are all quintessential novel ideas and practices in this part of the world (as they are undoubtedly to several other places). They also, importantly, did not grow out of local social realities and sensibilities but were externally imposed, as part of colonial and post-colonial states climbing the hills and attaching their institutions and policies to nearly all aspects of livelihoods and lifeworlds. On the ground, these borders, and the regulatory regimes they effectuate, more often than not conflict with historical and sociological common sense. Nowhere is this disjuncture—or to put it more bluntly, this very insanity—more evident than in the Konyak Naga village of Longwa where the Indo-Myanmar border runs through the house of its traditional *angh* (king), dividing his kitchen and bedroom between two countries, thereby turning the king's mundane everyday acts of brewing tea, retreating to bed or relieving his bowels into trans-border affairs.

In a classic essay, 'The Frontiers of "Burma"', Leach (1960) convincingly and eloquently showed how modern 'Western' political grammar and concepts, such as state, nation and border, are 'not necessarily applicable to all state-like political organisations everywhere'. They were certainly alien to the history and sociology of traditional Burma, an area Leach captured as 'the whole of the wide imprecise defined frontier region lying between India and China and having modern political Burma at its core' (Leach 1960, 5). Here I take this region as also counting in the whole of Naga-inhabited areas, in the sense of contiguous, connecting and interlocking tribal upland polities that existed prior to state enclosure. Leach himself, in his earlier *Political Systems of Highland Burma*, suggested that his model of societies oscillating between hierarchical chieftainship and egalitarian forms of social organization, and of interpenetrating political systems, might equally apply to the Nagas (Leach 1954, 291).

Observing the absence of straightforward borders in this region, Leach preferred the concept of 'frontier', which he approached not in terms of a hard and fixed political line but as 'a border zone through which cultures interpenetrate in a dynamic manner' (Leach 1960, 50). It was through this lens that Leach approached the social landscape, and in particular the complex dialectic between the 'valley' and 'hill people', or the Shan and Kachin, respectively. In his words: 'the indigenous political systems which existed prior to the colonial expansion were not separated from one another by frontiers in the modern sense and they were not sovereign nation-states' (Leach 1960, 50). The problem with historiography that adopted a state framework, he argued, was that such accounts were guided by the 'European myth' of absolute and indivisible sovereignty, thence wrongly relying on an ideology and political principle that sees all states as sovereign over a delimited piece of the earth's surface in such a way that there can be no overlay between territories.

In Burma, to the contrary, Leach (1960, 50) observed how 'the frontiers which separated the petty political units within "Burma" were not clearly defined lines but zones of mutual interest'. These political units, further, 'had interpenetrating political systems, they were not separate countries inhabited by distinct populations'. This notion of frontier as a 'border zone' through which cultures and polities interpenetrate and mingle in a highly dynamic manner, Leach emphasized, needs to be distinguished from what he termed 'the precise McMahon lines of modern political geography'. Such zones of 'mutual interest' have been widely documented not only in the hills but also in the adjoining plains of Assam where colonial officers, as they extended their sway into the valley, had to 'contend with the conditions of overlapping territoriality and sovereignty, which characterized the indigenous polity of this region' (Misra 2005, 22). It also extended between the plains and the foothills, with Naga upland villagers exerting negotiated access and measures of control over certain territories in the plains which were inhabited by Assamese (Wouters 2011).

These 'zones of mutual interest' and 'interpenetrating political systems' became disrupted with the definite state enclosure of the hills and the rise to hegemony of the nation state as the near-exclusive political form of viewing and managing territories. Scott (2009, 11) writes thus:

> State power, in this conception, is the state's monopoly of coercive force that must, in principle, be fully projected to the very edge of its territory, where it meets, again in principle, another sovereign power projecting its command to its own adjacent frontier. Gone, in principle, are the large areas of no sovereignty or mutually cancelling weak sovereignties. Gone, too, of course, are peoples under no particular sovereignty.

Various factors were considered in the subsequent drawing of political borders (done in South Asia mostly from drawing rooms in London) and in the demarcation and identification of national territories. Ethnicity, however, was not prime among these, and the section that follows shows how the enactment of the Indo–Burmese border resulted in a local mismatch between state spaces and the Naga ethnoscape.

STATE VERSUS ETHNIC SPACES

In the author's note to the novel *Remains of Spring: A Naga Village in the No Man's Land,* Jibon Krishna Goswami, a former member of an armed revolutionary group in Assam, reflects:

> The No Man's Land is a political geography, created due to the formation of two nations, India and Burma (now Myanmar). However, people had been living in this region since ages. It has been their land, socially and historically, and not the No Man's Land. They carried that history in their hearts and minds. Unfortunately, that history was maintained orally. It could not withstand the sword of colonialism, which bifurcated them on the day the two nations were born. In reality, three nations were born that day, one without a land of its own. It was the No Man's Land. This land became a world of its own. Its people lived on either side of the border.... (Goswami 2016, x)

This 'No Man's Land' broadly corresponds to A. Z. Phizo's depiction of the 'untouched Nagas' he introduced as he crafted his thesis for Naga independence in the 1950s. He wrote:

> The Nagas were divided by the British administration into three major units. About one fifth of the Naga population with that much in proportion of our land were administrated from British India. About the same proportion as administrated by British Burma. And approximately sixty percent of the population occupying a territory of about seventy percent of Nagaland [Naga lands] were left untouched and undisturbed, who were absolutely independent. (cited in Nuh 1986, 101)

In his commentary on Goswami's novel, Baruah (2016, xxiv) writes: 'The No Man's Land is shown as an irreducible irrationality of the coming into existence of postcolonial India and Burma with their respective boundaries, a process that violently affected the Nagas without their having any say in it.' Today, this 'No Man's Land' has by and large ceased to exist, as, over the past decades, the Indian and Burmese governments have projected their territorial powers and jurisdiction right up to the international border, so (nearly) cancelling out the possibility of stateless territories. As Goswami himself reflects: 'The reference to the area as the No Man's Land has seen a gradual decline. However, since the context of the story is set in the 1990s, the term 'No Man's Land' has been deemed appropriate' (Goswami 2016, vii).

The Indo-Burmese border was formally demarcated after the withdrawal of the British Raj. At first, it merely existed on the political map but did not correspond with state institutions, military surveillance or even boundary stones on the ground. Instead, this area became the habitat of Naga underground groups (and not only Naga) that formally objected to any such border within what they framed as ancestral and cultural Naga territory. Over time, however, this 'free zone' became progressively smaller and, barring a few patches here and there, has now been obliterated as the result of India and Myanmar—akin to all modern states—insisting that territorial sovereignty should be 'fully, flatly and evenly operative over each square centimetre of a legally demarcated country' (Anderson 2007, 19). Also akin to the idea of modern statehood, India and Burma became 'increasingly committed to, and reliant upon, their ability to make strict demarcations between mutually distinct bodies of citizens, as well as among different groups of their own subjects' (Torpey 2000, 12). This strict demarcation, including the patrolling of the political boundary, now contradicts, and complicates, the social realities and relations of Naga villagers, who, as a state effect, became borderlanders. It also contradicts, as noted, Naga nationalist projections of a unified Nagalim, because of which this border demarcation remains a volatile issue, as I will presently illustrate.

When, in 2016, the Indian and Burmese governments began fencing and trenching the international boundary, Naga organizations immediately objected. A spokesperson of the 'trans-border' Khiamniungan Naga—I am placing 'trans-border' within hyphens because the Khiamniungan resided there long before the international boundary was drawn—condemned the fencing as a 'felonious act'. They avowed that the 'Khiamniungan have always lived as one community', but that 'after British colonialism an imaginary line was drawn between India and Myanmar, dividing the tribe between two countries'. While colonial officers were extensively engaged in surveying Naga territory and making maps and developed an invented sense of where British India ended and British Burma began, this did not, in those days, culminate into a precise political boundary within the Naga area. Note, for instance, Hutton's (1987, iii) depiction of the Chang Naga:

> The Chang is one of those Naga tribes which occupy the hinterland, as it were, of the Naga Hills District, stretching back to the high range, which divides Assam from Burma. Only two small Chang villages of mixed population fall far enough west to come within the boundary of the administered district, the bulk of the tribe being situated in the area of loose political control which forms a buffer between the district and the still unknown tribes which occupy the slopes of the high range on both the Assam and Burma sides.

The boundary, back then, was imagined to be somewhere 'to the high range', but British officers refrained from narrowing it down to an exact line. In concrete form, the boundary line was delineated only after the withdrawal of the British Empire, and popular history has it that it was 'drawn over the Patkai ranges when Jawaharlal Nehru, the then Prime Minister of India, and U Nu, the then Prime Minister of Burma, flew over the area to determine the international boundary, thus unwittingly dividing villages perched on the mountaintops between the two nations' (Joshi 2013, 166). While loosely, if at all, patrolled at first, over time the inviolability of this border turned dogmatic, and so permanently divided not just the Khiamniungan but also the Konyak, Pochury, Tangkhul and other Naga tribes, as well as the projected Naga nation more widely. It is the redrawing of these borders—both international and national—that has since lain at the heart of the Naga struggle.

It is in this context that, when the fencing of the international boundary began, the NSCN-IM released a public statement in which it objected to the physical demarcation of a border in 'the heart of the Naga homeland', then warning: 'We shall no longer accept any policy to further divide the Naga family in the form of an artificial boundary fencing between India and Myanmar'. The Naga Hoho—a pan-Naga apex body—similarly rejected the border fencing and called it an 'attempt by India and Myanmar to rewrite the history of Nagas'. Nagaland's then Chief Minister Shurhozelie Liezietsu, in turn, stated: 'Even if we may not be able to do much to change the international boundary, we will do everything to see that the traditional rights of the Naga people to move about freely within their own ancestral land is not taken away'. These public outpourings were followed up by Nagaland's lone representative at the Lok Sabha writing, in a letter addressed to the prime minister, that 'the people living in both sides of the border in the Naga areas belong to similar tribes and have been living as one community since time immemorial' (Wouters 2018, 11–13).

It is in an attempt to recognize and institutionalize this long-standing connection of Nagas currently residing on either side of the border that the following, and final, section explores the possibility and idea of 'supranational citizenship'.

SUPRANATIONAL NAGA CITIZENSHIP

After decades of violent struggle, countless victims and long years of slow, painful political negotiations, it increasingly appears that a permanent Indo-Naga political settlement—one that is satisfactory to all parties involved—would not be possible in terms of the modern and modular categories of the nation states, hard political borders and exclusive territorial sovereignty. This is because options and possibilities on the table of nation states often limit themselves to zero-sum games in which the gains of one party imply equivalent losses for another. In broad strokes, this seemingly explains the continuing abeyance of a now-long-awaited final political settlement. Given this apparent deadlock, it may be opportune to rethink and re-frame our political vocabulary (a unique, extra-constitutional arrangement, after all, may require a new political language), which, in turn, may add fresh political possibilities to the table. In this, some options may be found in the past—a return, that is, to Leach's notions of 'frontier', 'zones of mutual interest' and interpenetrating polities. As Baruah (2016, xlvi) aptly observes:

> If the entrapment of the Naga nationalist struggle in a perpetual state of being is a condition created by the existing nation states, then how would the Naga struggle emerge out of the language of nationality, which itself produced its condition of perpetuity in the first place?

This need to emerge out of the language of nationality is complex but crucial. The need to think beyond conventional political concepts and dogmas has now indeed been recognized. If newspaper reports and occasional statements by those involved are anything to go by, the ideas such as shared sovereignty, a supra-state body and non-territorial integration are currently being discussed and explored. Here, I propose to add another ingredient to this new political dish that is being prepared, which is that of 'supranational citizenship'. The term 'supranational', in the context of citizenship, may here be loosely defined as a form of institutionalized belonging and a set of rights and duties that extend beyond the political imagination and limitations of the nation state in such a way that it recognizes that ethnic and territorial belonging transcends existing political boundaries. On the face of it, supranational citizenship may seem a contradiction, as the concept of citizenship is inextricably linked to the nation state, even as some states allow for dual or multiple citizenship. Forms of supranational citizenship, however, are already a reality in some parts of the world, most prominently in the European Union (EU). Strumia (2017, 671) explains supranational citizenship in the EU thus:

> It denotes a status that stems from nationality, yet re-articulates citizenship beyond its boundaries. This re-articulation relies on a rule of mutual recognition: each member state in a supranational community recognizes national citizens of other member states to some extent as its own. Political and residence rights granted to supranational citizens in national communities beyond their own give concrete expression to this rule of recognition.

What I have in mind here is not something as grand and ambitious as European citizenship that wields people of many nations together, but to explore ways to recognize and institutionalize the Nagas as a supranational community, with both the Indian and Burmese state recognizing Nagas on the other side of the border 'to some extent as its own'. Exploring this necessitates a different understanding of the relationship between citizenship and nationality, one that acknowledges forms of transnational belonging. From this perspective, national citizenship becomes complemented by, so to speak, 'ethnic citizenship', thus asking both the Indian and Burmese states to bequeath certain rights on those Nagas who are not their national citizens but who nevertheless identify with territory and people on either side of the Indo-Myanmar border.

While the idea of an ethno-territorial homeland is no doubt fraught with dangers of its own (Van Schendel 2011), this form of 'adjacent' or 'supplementary' citizenship would nevertheless work to acknowledge the existence of a contiguous Naga homeland and so without formally redrawing existing political boundaries. It could additionally permit Naga civic and national rituals, production and circulation of Naga personal documents and pan-Naga development and welfare measures, as well as the formal enactment of a pan-Naga body (transcending both countries) that could exist alongside—and coordinating with—existing legislative assemblies. This body (of which the current Naga Hoho may be seen as a forerunner)—which may receive funds from both the Indian and Burmese states—would represent all Nagas, oversee their cultural, political and economic progress and may also levy taxes locally to be used for the common good; to be noted here is that within India, Nagas are currently exempted from paying income tax. As such, a Naga tax system would not amount to double taxation.

Besides explicit recognition, the institutionalization of a supranational Naga community may also be operationalized in the form of certain rights that apply to Nagas equally on either side of the border. These rights, to be sure, must be seen as supplementary, not as substituting the rights and obligations that already exist in relation to either Indian or Burmese citizenship. I briefly (and, admittedly, simplistically) invoke four such possible rights here: the rights of residence, employment, education and movement. Let us start with residence. This would entail that any Naga would be allowed to settle anywhere within Naga territory and procure land (in accordance with Naga customary law), even if this would mean a shift from India to Myanmar, or vice versa.

The right to employment, here, would mean that jobs—both in the local government and in private offices—would be opened for Nagas on both sides of the border. Next, the right of education would enable students from across Naga territory to seek admission in any educational institution within Naga territory, with all such institutions functioning under a single Naga school board that is equally recognized by the Indian and Burmese states.

Finally, the right of movement would allow Nagas to travel, trade and move around freely within Naga territory, without being held up or checked at the international border. In a limited way, this policy is already in the making. 'Villagers on India-Myanmar border to get passes' (Singh 2018). The article then explained that border passes would soon be given to those living in the vicinity of the border and that such a pass allows its bearer to cross the border and move freely, without needing any permit or visa, up to 16 km into either Indian or Burmese territory. An official explained: 'The border pass will be given only to the domiciles. All residents going across the border for agriculture, work or to meet relatives should carry the pass at all times.' This may be interpreted as a conservative beginning to what supranational Naga citizenship would enable: the free and unchecked movement by Nagas across the length and breadth of the Naga ethnoscape.

Each of these possible pan-Naga rights invoked here would come with complications and difficulties in their implementation, and they are here mentioned only as preliminary ideas. However, regardless of any such practical difficulty, what the theory and praxis of recognizing and institutionalizing the Naga as a supranational community with specific rights would crucially accomplish is, first, legitimizing the enduring sense of Naga identity and belonging which refuses to be limited to existing political boundaries and, second, offering Nagas on either side of the border the opportunity to pursue shared economic, political and cultural goals.

CONCLUSION

If the case of the Garos—as introduced in the beginning of this chapter—stands out as a clear example in which the enactment of national boundaries and territorial sovereignty largely succeeded in disconnecting and obscuring pre-existing regional ethnic tribal unity and, over time, worked to localize economic aspirations and political claims within different national spaces, the Naga movement continues to reject such curtailment of ethnic identification, belonging and common aspirations. After briefly discussing the historicity of the Indo-Myanmar border, and the disjuncture between state and ethnic spaces that its demarcation produced, this chapter invoked the idea of supranational citizenship as a possible way to overcome current political and territorial divisions. It is proposed that supranational citizenship—both in recognition and as a form of praxis—denotes a way to bypass the dogma of the inviolability of the border through institutionalizing the spatial and sociopolitical organization of the Nagas in a way that does not corrode or undermine existing political and territorial arrangements.

Operationalizing supranational citizenship is no doubt fraught with complexities and controversies, and it can only succeed when both the Indian and Burmese states are willing to engage in political gymnastics (and also if different Naga factions smoothen their differences). What remains evident, though, is that a permanent political settlement may only be possible through a new set of political concepts of practices, of which supranational citizenship may be one among several others.

REFERENCES

Abraham, Itty, and Willem van Schendel, eds. 2005. *Illicit Flows and Criminal things: States, Borders and the Other Side of Globalisation.* Bloomington: Indiana University Press.

Anderson, Benedict. 2007. *Imagined Communities: Reflections on the Origins and Spread of Nationalism.* London: Verso.

Ao, L. 2002. *From Phizo to Muivah: The Naga National Question in Northeast India.* New Delhi: Mittal Publications.

Bal, Ellen, and Timour Claquin Chumbugong. 2014. 'The Borders that Divide, the Borders that Unite: (Re)interpreting Garo Processes of Identification in India and Bangladesh'. *Journal of Borderlands Studies* 29(1):95–109.

Banerjee, M., and R. P. Athparia. 2004. 'Emergent Ethnic Crisis: A Study of Naga-Kuki Conflict in Manipur'. *The Journal of the Anthropological Survey of India* 53(1):77–90.

Baruah, Manjeet. 2016. 'Introduction. Villages and maps: Space, text, and people.' In *Remains of spring: A Naga village in the No Man's Land*, edited by Majeet Baruah, xxii–li. Delhi: Oxford University Press.

Bourdieu, Pierre. 1993. *Sociology in Question*. Translated by Richard Nice. London: SAGE.

Chasie, C. 2005. *The Naga Imbroglio: A Personal Perspective*. Kohima: City Press.

Gellner, David N. 2012. Uncomfortable Antinomies: Going beyond Methodological Nationalism in Social and Cultural Anthropology. *AAS Working Papers in Social Anthropology* 24:1–16.

Gellner, David. 2013. 'Introduction: Northern south Asia's Diverse Borders, from Kachchh to Mizoram.' In *Borderland Lives in South Asia,* edited by David Gellner, 1–23. Durham and London: Duke University Press.

Goswami, Jibon Krishna. 2016. *Remains of Spring: A Naga Village in the No Man's Land* (translated by Majeet Baruah). Delhi: Oxford University Press.

Horam, Mashangthei. 1975. *Naga Polity*. Delhi: D.K. Fine Art Press.

Hutton, John H. 1987. *Chang Language, Grammar, and Vocabulary*. New Delhi: Gian Publishing House.

Joshi, Vibha. 2013. 'The Micropolitics of Borders: The Issue of Greater Nagaland (or Nagalim).' In *Borderland lives in Northern South Asia,* edited by David N. Gellner, 163–193. Durham and London: Duke University Press.

Leach, Edmund. 1954. *Political Systems of Highland Burma: A Study of Kachin Social Structure*. Cambridge, MA: Harvard University Press.

Leach, Edmund, R. 1960. 'The Frontiers of "Burma"'. *Comparative Studies in Society and History* 3(1):49–68.

Lintner, Bertil. 2011 [1990]. *Land of Jade: A Journey from India through Northern Burma to China*. Bangkok: Orchid Press.

Misra, Sanghamitra. 2005. 'Changing Frontiers and Spaces: The Colonial State in Nineteenth-Century Goalpara'. *Studies in History* 21(2):215–246.

Nuh, V. K. 1986. *Nagaland Church and Politics*. Kohima: V. Nuh & Bro.

———. 2002. *My Native Country. The Land of the Nagas*. Guwahati: Spectrum Publications.

———. 2006. *165 Years of Naga Baptist Churches*. Kohima: Council of Naga Baptist Churches.

Scott, James C. 2009. *The Art of not Being Governed: An Anarchist History of Upland Southeast Asia*. New Haven and London: Yale University Press

Singh, Vijaita. 2018. Villagers on India–Myanmar Border to Get Passes. 13 January. *The Hindu*. https://www.thehindu.com/news/national/villagers-on-india-myanmar-border-to-get-passes/article22437211.ece (accessed 7 February 2021).

Strumia, Francesca. 2017. 'Supranational Citizenship.' In *The Oxford Handbook of Citizenship,* edited by Ayelet Shachar, Rainer Bauböck, Irene Bloemraad, and Maarten Vink, 669–693. Oxford: Oxford University Press.

Torpey, John. C. 2000. *The Invention of the Passport: Surveillance, Citizenship and the State*. Cambridge: Cambridge University Press.

Van Schendel, W. 2005. *The Bengal Borderland: Beyond State and Nation in South Asia*. London: Anthem Press.

Van Schendel, W. 2011. 'The Dangers of Belonging.' In *The Politics of Belonging in India: Becoming Adivasi*, edited by D. Rycroft and S. Dasgupta, 19–43. London: Routledge.

Vashum, R. 2005. *Nagas' Right to Self-Determination: An Anthropological-Historical Perspective*. New Delhi: Mittal Publications.

Wouters, Jelle J. P. 2011. 'Keeping the Hill Tribes at Bay: A Critique from India's Northeast of James C. Scott's Paradigm of State Evasion'. *European Bulletin of Himalayan Research* 39:41–65.

———. 2016. 'Sovereignty, Integration or Bifurcation? Troubled Histories, Contentious Territories and the Political Horizons of the Long Lingering Naga Movement.' *Studies in History*, 32(1):97–116.

———. 2018. *In the shadows of Naga insurgency: Tribes, State, and Violence in Northeast India*. Delhi: Oxford University Press.

Chapter 18

Chin Migration to Mizoram
Ethnic Affinity and Changing Perceptions

R. K. Satapathy and P. C. Lalthansiami

INTRODUCTION

Population movement is as old as human civilization. Throughout history, people have migrated from one place to another owing to natural catastrophes or calamities, upheavals, economic compulsion, political insecurity and human rights violations (Debbarma and George 1993). There is no universal agreement on the definition and conceptualization of migration. In its more general sense, migration refers to permanent or semi-permanent movement of persons over a significant distance. In recent years, because of the complexity of human life and fast-changing socio-economic conditions, human migrations have increased in number and in magnitude. There are about 272 million international migrants around the world; the majority migrated in search of better economic and social opportunities. Others are forced to flee by war, conflict, generalized violence and human rights violations. Movement of population from one geographic location to another has varied consequences, which can either be positive or negative. The impact can be felt both at the place of origin and at the place of destination. The cost of migration is greater on the recipient country than on the place of origin of the migrants and can affect economic structures, the environment, demography and culture. Further, large-scale migration can cause various epidemic diseases, socio-economic problems, political and social conflicts or tensions, cultural and linguistic conflicts, etc., between the indigenous people and the migrants (Debbarma 1993, 237–238).

India has been a host to millions of migrants who came in search of safety, security and development. Thousands of migrants came from Myanmar, Tibet, Bangladesh, Sri Lanka, Afghanistan, Pakistan, Nepal and Bhutan and settled in India. Different states in Northeast India have also witnessed people coming from outside their region. The most prominent migrants in the North Eastern Region are Tibetans, Nepalis, Bangladeshis, Chakma, Brus, Chin and Bengalis. The continued influx of migrants has been of deep concern in the region, as the increase in the growth rate of the population creates several socio–economic and political problems in the region.

MIGRATION OF CHIN TO MIZORAM

As one of the eight states in the North Eastern Region of India, Mizoram has experienced migration of people from its neighbouring states such as Assam, Tripura and Manipur. It has also received international migrants from Nepal, Bangladesh and Myanmar. Unlike other migration, the trans-border migration of Chin from Myanmar to Mizoram is viewed differently, as the people of Myanmar's Chin Hills, often referred to as 'Chins', and the people of Mizoram, who identify themselves as 'Mizos', belong to same ethnic origin. Although the Chin and the Mizos are now separated by the international border, most of the Mizos in Mizoram and Chin in Chin State of Myanmar believe that they have a common ancestor who emerged into this world from the bowels of the earth or a cave or a rock. Most Zo tribes refer to it as Chhinlung; however, some Zo dialects have other names for the 'cave' or 'hole'. The Chin and Mizos have the same genetic, cultural, historical and linguistic heritage, and their dialects sound similar.

It was only during the colonial period that these agnate communities were given different names, as Lushai, Kukis and Chin. However, even during the colonial period, when the Indian subcontinent and

Burma were under the British, there were no restrictions on the movement of these tribal communities.[1] Even in the post-independence period, even after the border agreement between India and Burma, there were normal relations between the Chin of the then Burma and the Mizos for over a decade. However, because of the political crisis, followed by ethnic oppression and religious discrimination in Burma, the Chin began crossing the border and entered the Lushai Hills, now Mizoram. The waves of migration which started in the late 1960s gathered momentum after the 1988 pro-democratic uprising in Myanmar. Many Chin fled to India to escape brutal military crackdown, and many people fled to Mizoram to escape arbitrary arrest, torture, forced labour and religious persecution. The 1988 Uprising against the Burmese military government left thousands of citizens dead in many parts of the country. Chin State was heavily militarized, and the poverty and military oppression that they had to endure compelled many Chin to cross the border and take shelter in India's Mizoram state (UNHCR 2019). During the initial period of the migration to Mizoram, India's union government and the government of Mizoram provided refuge to the Chin, giving them aid and assistance in Champhai and Saiha districts of Mizoram. However, these services were stopped on 1 June 1995.[2] The Chin migrants continued to stay in Mizoram due to continued conflict, persecution and economic insecurity in Myanmar, and their numbers in Mizoram have grown to over 1 lakh (Central Young Mizo Association 2011). Although the majority of the Chin migrants still live in such border districts as Champhai, Saiha and Lawngtlai, over the years they have spread to all parts of Mizoram.

FACTORS CONTRIBUTING TOWARDS CHIN MIGRATION TO MIZORAM

Chin State, created in Myanmar in 1974, remains impoverished and underdeveloped. The Tatmadaw (official name of the armed forces of Myanmar) in Chin State imposed several restrictions on the ethnic, civil, political and religious rights of the Chin. Tensions with the Tatmadaw increased after the 8-8-88 Uprising. There has been an increase in the reported cases of forced labour, human rights abuse, ethnic and religious persecution, torture, sexual violence, forced conscription into the armed forces, summary killings and arbitrary arrests

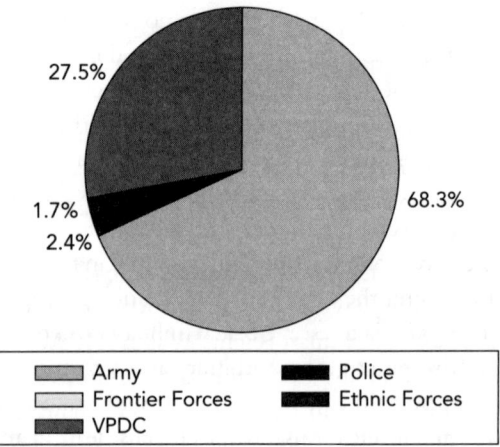

Figure 18.1 *Proportion of Crimes Against Humanity by Alleged Perpetrators in Chin State*

Source: Sollom et al. (2011).

against local Chin by the Tatmadaw. A report, 'Health and Human Rights in Chin State, Western Burma: A Population-Based Assessment Using Multistage Household Cluster Sampling', detailed the crimes against humanity as shown in Figure 18.1. According to the report, 91.9 per cent of the households in Chin State were victims of direct physical violence, including killing by guns or other deadly weapons, beating, torture, sexual violence, forced conscription into the armed forces, human rights abuses, including food security–related abuses, forced labour, disappearance and ethnic and religious persecution. Of the abuses in Chin State, 63.3 per cent were committed by the army. Village Peace and Developmental Councils (VPDCs) were responsible for 27.5 per cent of severe-abuse cases. The police accounted for 4.9 per cent and ethnic forces for 1.7 per cent of the abuse cases (Sollom et al. 2011).

The maximum number of people fled from Chin State because of economic insecurity in Myanmar. While obstacles to daily survival, including earning a livelihood, are reasons for the continuing exodus from Chin State to Mizoram, the economic situation of the Chin can only be understood within the context of the multiple persistent human rights violations being committed against the Chin by the Tatmadaw. The government of Myanmar allocated 24 per cent of its annual budget for the army and spent only 4.3 per cent on education and 1.3 per cent on health (Thang 2011). A survey conducted by the United Nations Development Programme (UNDP) showed that Chin State is the poorest among all the states in Myanmar and that 73 per cent of Chin live below the poverty line. In addition, around 27 per cent of Chin children have no access to primary school, and 32 per cent of Chin people do not have access to healthcare (IHLCA Project Technical Unit 2011). Human Rights Watch (HRW) reported that there are only 12 hospitals, 56 doctors and 128 nurses in Chin State. The quality of healthcare in Chin State is poor, and medical treatment is costly. The quality of the education system is also bad in Chin State, where there are 1,167 primary schools, 83 middle schools and 25 high schools, and there are no universities in the region (Human Rights Watch 2009). Thus, decades of neglect, underdevelopment and widespread abuse have devastated the Chin and compelled many to migrate to and find refuge in Mizoram.

The Chin have continued fleeing from Chin State also because of religious persecution. The Rohingya Muslims in Rakhine State are not the only ones to face religious discrimination in Myanmar. The Chin, for a long time, have been the victims of religious discrimination and persecution in Myanmar. The government, biased towards Buddhism, the official religion in Myanmar, looks at the Chin with suspicion. The Chin, the majority of whom are Christians, are discriminated against, tortured and jailed because of their faith. In fact, the Refugee Council USA reported that Myanmar is among 'the countries of particular concern' for its severe restrictions on religious freedom.[3] The Chin Human Rights Organization (CHRO) reported in October 2018 that the Chin Christians have faced attacks and violence from the majority population, the local police force and resident monks who have been catalysts for violence. This has manifested more violently in areas where Buddhists and Christians live in proximity, such as the Chin State and Kachin State in Myanmar (Chin Human Rights Organization 2018).

In addition, Myanmar has been afflicted by ethnic conflict and civil war since its independence in 1948, exposing it to some of the longest-running armed conflicts in the world. The situation worsened for ethnic minority groups after the military coup in 1962, when their rights were further curtailed. The military government has as yet refused to concede to political demands of ethnic minority groups, mostly treating ethnic issues as military and security issues (The Transnational Institute). Consequently, ethnic-minority communities, such as the Chin, have long felt marginalized and discriminated against. The failure of the government to protect and assist its minority communities, the fight between smaller ethnic groups and the insurgency and counter-insurgency movements are also responsible for the flight of the Chin to Mizoram. On 18 January 2019, the United Nations (UN) Special Rapporteur on Human Rights 'expressed alarm at the escalating violence in northern and central Rakhine State and Chin State' and further stated that that 'since November 2018, the Myanmar military, and Arakan Army, an ethnic armed organization, have been engaged in heavy fighting in the Chin State, resulting in deaths and injuries to civilians. At least 5,000 people have been displaced from their homes' (UNOHCHR 2019).

Migration of the Chin to Mizoram is facilitated by both push and pull factors. While the factors that push the Chin out of Myanmar include political instability, human rights violations, religious persecution, economic underdevelopment, food insecurity, poverty and ethnic violence in Myanmar, the factors that pull the Chin towards Mizoram include better economic opportunities, religious freedom, better education system and stable political system. In addition, the ethnic relationship between the Chin and Mizos also plays an important role in the migration of the Chin to Mizoram.

SOCIOPOLITICAL CONSEQUENCES OF CHIN MIGRATION TO MIZORAM

Mizoram shares its border with Myanmar and Bangladesh and the Indian states of Assam, Manipur and Tripura. Apart from the indigenous tribes, the population of Mizoram includes migrant workers from mainland India and also undocumented and illegal migrants from Bangladesh and Myanmar. The attitude of the local Mizo population towards these migrants is different. Not only the illegal migrants from Bangladesh but even the migrants from mainland India are not considered as genuine people of Mizoram; instead, they are viewed as outsiders, as *vai* in the local dialect. Only the Chin from Myanmar are treated better, owing to ethnic commonality. Although the Chin are undocumented migrants from Myanmar, they are often able to acquire documentation associated with citizenship, such as a valid voter identity (ID) card, ration card or passport in Mizoram. Acquisition of such documents becomes possible because of the help and cooperation from the host population. The Chin migrants received support and cooperation from Mizo voluntary organizations and the local Mizo population. Apart from ethnic affinity, the Chin and Mizos also share the same religion (Christianity), and hence the Chin have more advantages compared to the other non-Christian migrants in Mizoram. There is no official record about the number of Chin refugees in Mizoram. Neither the state government nor local authorities nor the non-governmental organizations (NGOs) have any accurate data about the number of Chin migrants in Mizoram. It is very difficult to know their exact number because they are illegal migrants, and they have easily mingled with the local population because of their common ethnic roots and linguistic similarity. They are employed primarily in the unorganized sector as agricultural labourers, domestic helpers and artisans. Some of them own lands and even secure government jobs. Although their economic status is not equal to that of the Mizos, the Chin are better off than other migrants who have settled in Mizoram. Though they are not recognized as refugees, they are more secure compared to other non-Mizo communities (Sadiq 2019, 36).

Initially, the ethnic commonalities and a sense of brotherhood compelled the Mizos to be sympathetic towards the plight of the Chin in Mizoram. However, as the population of the Chin continued to increase, the attitude of the government, NGOs and the Mizo civil society towards the Chin migrants began to change. In recent decades, Mizoram has witnessed periodic 'anti-foreigner' campaigns that eventually led to large-scale deportation of Chin to Myanmar. During these campaigns targeting the Chin, the Mizo public, NGOs and state authorities have threatened the Chin migrants with arrest, forced eviction from their homes and deportation to Myanmar.

The first case of large-scale Chin deportation from Mizoram took place on 16 July 1996 when C. Lalhmangaiha, the president of the Khuang Leng village unit of the Young Mizo Association (YMA), was allegedly shot dead by three members of the Chin National Army (CNA). This created tension between the Chin and the local Mizos, and thousands of Chin, who are believed to be supporters and sympathizers of the CNA/Chin National Front (CNF) were arrested, jailed and deported back to Myanmar by the Mizo authorities and Mizo voluntary organizations.[4] The second case of the Chin's deportation from Mizoram took place on 17 July 2003, when a minor Mizo girl was raped by a Chin boy in Aizawl. After this incident, the local Mizo NGOs, supported by the police and political parties of Mizoram, started a massive eviction programme against Chin migrants in Mizoram.[5]

As the population of Chin migrants increases over the years, they are viewed as a burden on the Mizo society. To check the growing infiltration, several campaigns were organized by the local NGOs in Mizoram. On 25 July 2003, leaders of the prominent Mizo NGOs such as Central YMA, Mizo Zirlai Pawl (MZP), Mizo Upa Pawl (MUP) and Mizo Hmeichhe Insuihkhawm Pawl (MHIP) submitted a joint memorandum to the then chief minister of Mizoram, urging the government of Mizoram to place a check-gate at the Myanmar–India border, so that illegal immigration in Mizoram and illegal smuggling of drugs, alcohol and other items could be monitored and checked.[6] In response, on 27 March 2004, the government of Mizoram enforced certain rules and regulations for entry at the India–Myanmar border of Mizoram. The Chin are permitted free movement for up to 16 km in Indian territory without valid travel documents.[7] As per the rules, Burmese nationals coming from across the border through Zokhawthar, Champhai district, were charged ₹100 as entry fee. The temporary permit for staying in Mizoram is priced at ₹20, and the application form is priced at ₹20 per unit. Burmese migrants are not allowed to stay in Mizoram for more than 30 days continuously, and they can extend or renew their permit through the District Superintendent of Police by paying a renewal fee of ₹100 per head. However, under no circumstance are Burmese nationals permitted to stay in Mizoram for over 8 months. There are two checkpoints, one on Mizoram's side and the other on Myanmar's side. The Mizoram entrance checkpoint at the Iron Bridge in Zokhawthar–Rih is managed by the Mizoram Police, and on the Myanmar side the gate is manned by Myanmar customs and immigration personnel. The gate is kept open from 7:00 a.m. and is closed at 5:00 p.m.[8] However, most Chin hardly follow these rules, and one can find many unregistered Chin in different districts of Mizoram. On July 2017, the Ministry of Home Affairs (MHA) announced a plan to systematize the Free Movement Regime (FMR), claiming that the FRM, which allows the Mizo people living in Myanmar to enter Mizoram up to 16 km, has been misused by terrorists and smugglers. Additional Secretary for Home Lalbiakzama added that the MHA expressed concern over the abuse of the FMR, as insurgents from both the Indian and Burmese sides used it to move freely across the porous border areas. He pointed out that apart from the militants, smugglers, drug traffickers and gunrunners were also using the same routes (NDTV 2017).

To check illegal infiltration from Myanmar, on 27 September 2017, the union government opened the Zorinpui land checkpoint in Lawngtlai district of Mizoram. The remote Zorinpui village is located 287 km away from Sittwe Port in Myanmar (*Times of India* 2017). On 30 August 2018, an immigration checkpoint was also opened at Zokhawthar in Champhai district of Mizoram (Business Standars 2018). Despite these measures, the Chin continued to cross the Indo-Myanmar border through the main Zokhawthar–Rih checkpoint and reach Aizawl and other parts of the state of Mizoram. Simultaneously, the increased transportation facilities also enabled the Chin people living across the border to quickly move back and forth across the Indo-Myanmar border through Champhai—the Zokhawthar–Rih sector.

The inability to check the flow of the Chin made the Mizo community view the overwhelming presence of the Chin in the state as a threat to the economic, political and societal stability of Mizoram. The opposition to Chin's presence in Mizoram is largely based on claims that they are responsible for alcohol brewing and drug trafficking. The Chin in Mizoram are also accused of the crimes such as robbery, rape and murder of the locals. J. Pachuau stated, 'all crime in Mizoram, from petty theft to gruesome murders, was blamed on the Burmese' (Pachuau

2014, 193). Drug addiction has been a debilitating social problem in Mizoram for over 20 years. Several youths have become addicted to opium, heroin, amphetamines and methamphetamine. The meth factories in Myanmar are partly fuelled by pseudoephedrine smuggled out of India, mainly through the Northeast. Mizoram, which shares a 404-km unfenced border with Myanmar and is in the neighbourhood of the Golden Triangle—an area that overlaps the mountains of Myanmar, Laos and Thailand—serves as a two-way route for cross-border drug trade based in Myanmar. An Excise and Narcotics Department official stated that 'Pseudoephedrine is bought from other parts of India and smuggled out to Myanmar through Mizoram, which is then used to make methamphetamine and other derivatives which is smuggled across the world' (*Telegraph* 2014). Admitting that there has been drug trafficking along the India–Myanmar border, the Superintendent of Police in Aizawl district, L. R. Dingliana Sailo, stated that both migrants and locals share the blame. He emphasized, 'There is a need to increase anti-poverty work, so refugees and locals are not tempted by illegal activities' (*Telegraph* 2014). However, the Mizo NGOs blame only the Chin for these drug-related activities. Lalhmachhuana, the president of the MZP, said: 'So many refugees are engaged in this trafficking... we need to implement good laws and we need better control (of the border)' (UNHCR 2014, 2019). Blaming them for not abiding by the rules and regulations of the host society, on 15 April 2013, YMA issued a quit-Mizoram notice to the Chin settlers in Phunchawng and Rangvamual villages on the outskirts of Aizawl.[9] Again, in August 2016, YMA, MUP and the Mizo Joint Action Committee (MJAC) demanded the removal of illegal Chin from Mual Khang village who were alleged to have been brewing alcohol.[10] Responding to the allegation that the livelihood of the poor people was being taken away, YMA asserted that 'there are other ways of earning and living' (Mizo Archive 2016).

It is true that many Chin migrants are involved in a number of illegal activities, such as stealing, drug peddling, alcohol brewing, etc. However, there is no evidence to prove that the Chin migrants in Mizoram commit more crimes than the local Mizos. It is said that most of the drug dealers or sellers in Mizoram are Mizos; the Chin are made scapegoats for smuggling the drugs into Mizoram from the infamous

Golden Triangle where Myanmar, China and Laos meet. Similarly, the majority of the liquor factory owners in Mizoram are Mizos. They hire Chin for alcohol brewing. Still, the Chin are blamed for the criminal activities and are accused of having a bad and 'unChristian' influence on Mizo society.

This growing mistrust and antipathy towards the Chin should be seen in the context of the changing political economy of the state. The growing number of Chin migrants in Mizoram increases the state's population and creates an additional burden on the state's resources. The presence of a sizeable migrant population generates competition between the locals and the Chin for control over the resources. The influx of cheap manufactured goods from the border areas such as Zokhawthar, Vaphai, Tlangsam, Para, Hlung Mang, Chapui–Matupi and Sangau–Lungpher makes it difficult for the local Mizos selling Indian goods to compete. The increasing involvement of the Chin in informal trade creates jealousy and resentment among the locals in the urban areas of Aizawl, and also in Champhai and Saiha districts.

The emphasis on the negative consequences often results in over-looking the positive effects of the Chin migration into Mizoram. It is primarily because of the Chin involved in formal and informal trade that there have been brisk trade transactions across the Myanmar–Mizoram border. Mizoram's prosperity is heavily dependent on trade through the Indo-Myanmar border, particularly in the Zokhawthar–Rih sector. In fact, every Mizo household depends on the cheap consumable and household products from the Indo-Myanmar trade. The Chin, known for their hard work, make cheap labour available to the locals and contribute to the development of infrastructure. It is easy for the local Mizos to deal with them, because they belong to the same ethnic community and speak in the language known to them. The Chin migrants in Mizoram have also contributed to the music and fine arts. The presence of the Chin contributed to the strengthening of the Zo identity movement in Mizoram.

Despite resentments against the Chin, the living conditions of the Chin migrants have improved over the years. In recent years, there has not been any serious attempt at deportation of the Chin. The growing knowledge of the sorrowful plight of the Chin in Myanmar and development of roads and communication helped the local Mizos empathize with the Chin of Myanmar. In 2015, the government of Mizoram, local NGOs, churches and the public in Mizoram sent aid and help to those affected by floods in Chin State (*Indian Express* 2015). The ethnic commonality and shared racial, religious, cultural and linguistic roots of the Mizos and the Chin were the prominent factors that motivated the Mizos to raise funds. The Chin, who were earlier referred to as foreigners (*Burma-mi*), are now described as 'our brothers and sisters in Burma'. The differences are overshadowed by the acknowledgement of the commonality of Mizos and Chin. This acknowledgement has come not from a fringe of local Mizos but by mainstream Mizos residing in Mizoram. The commonalities made it possible for the Chin to live in Mizoram despite differences and periodic tensions.

CONCLUSION

The Mizos of Mizoram and the Chin of Myanmar belong to the same ethnic family. It was this ethnic affinity that made the Mizos take a soft approach towards the Chin. The Mizos were up against the Bru and Chakma refugees, whom they considered as foreigners. However, initially when the Chin fled from Myanmar, the Mizos welcomed them and gave them shelter. The Chin took advantage of the open borders and entered Mizoram. Because of religious and linguistic similarities, it became easy for them to merge with the general population. Some Chin, with the help of the locals, could even acquire citizenship rights and started enjoying all facilities like other Mizos. The Chin provided cheap labour essential for the development of the state. However, as the population of the Chin began increasing due to the continuous flow of Chin into Mizoram, the Mizos started feeling the pinch. The government found it difficult to accommodate all the Chin migrants/refugees due to resource constraints. The Mizos started holding the Chin responsible for the drug menace and increasing criminal activities in the state. There was a gradual change in the public perception, leading to the demand for deportation of the illegal Chin migrants. This growing opposition to the growing number of

Chin showed the limitations of the ethnic affinity. Over time, the different historical trajectories of the Mizos and the Chin have led to the creation of separate regional/political identities and interests. No doubt, the Chin are preferred and tolerated more compared to other migrants and refugees having different ethnic affiliations. The Mizos no doubt sympathize with the Chin and walk an extra mile to help them when they face political or natural disasters. However, when the Mizos viewed their own interests as being threatened, they did not mind taking a stand against the Chin, despite the ethnic, religious and linguistic similarities. The uneasy relationship between the Mizos and the Chin shows the contradictory effects of colonial and post-colonial rule in the borderlands.

NOTES

1. Burma was part of British India till 1937 and later was the crown colony of Britain up to 1948.
2. Order No. 37 of the Champhai Sub-Divisional Office of the Government Of Mizoram, 1995.
3. A delegation of the Refugee Council USA went to different parts of Chin State of Myanmar and Mizoram state of India to study the plight of the Chin. For details, see, Refugee Council USA (2011).
4. Press release by the non-governmental organization (NGO) Coordination Committee, General Headquarters Aizawl, Mizoram, 28 July 1996.
5. Press release by Mizo Joint Action Committee, General Headquarters Aizawl, Mizoram, 20 July 2003.
6. A joint memorandum submitted after the joint meeting of all NGOs to the chief minister of Mizoram, 25 July 2003.
7. On 21 July 2010, through a gazette notification, the Ministry of Home Affairs, Government of India, reduced the benchmark of 40 km to 16 km.
8. *Vanglaini*, Aizawl, 30 March 2004, p. 1.
9. Press release, Young Mizo Association, General Headquarters Aizawl, Mizoram, 15 April 2013.
10. Press release, Young Mizo Association, Mizo Upa Pawl and Mizo Joint Action Committee (MJAC), General Headquarters Aizawl, Mizoram, 2 August 2016.

REFERENCES

Business Standars. 2018. *Land immigration centre along Myanmar border opened*. 30 August, Business Standars. https://www.business-standard.com/

article/pti-stories/land-immigration-centre-along-myanmar-border-opened-118083000881_1.html (accessed 30 September 2019).

Central Young Mizo Association. 2011. *Census of Burmese (Chin) Population in Mizoram*. Aizawl, Mizoram: YMA Headquarter.

Chin Human Rights Organization. 2018. *Stable and Secure? An Assessment on the Current Context of Human Rights for Chin People in Burma/Myanmar*. October. https://www.burmalink.org/stable-and-secure-an-assessment-on-the-current-context-of-human-rights-for-chin-people-in-myanmar-burma/ (accessed 10 August 2019).

Debbarma, P. K., and Sudhir Jacob George. 1993. *The Chakmas refugees in Tripura*. New Delhi: South Asian Publishers.

Human Rights Watch. 2009. *The Chin People of Burma: Unsafe in Burma, Unprotected in India*. New York: Human Rights Watch Press.

IHLCA Project Technical Unit. 2011. *Poverty Profile: Integrated Household Living Condition Survey in Myanmar 2009–2010*. Myanmar: UNDP. https://www.mm.undp.org/content/myanmar/en/home/library/poverty/publication_1.html (accessed 8 August 2019).

NDTV. 2017. *Free Movement Rules Along Mizoram-Myanmar Border To Be Tightened: Official*. 27 September. https://www.ndtv.com/india-news/free-movement-rules-along-mizoram-myanmar-border-to-be-tightened-official-1755915 (accessed 30 September 2019).

Pachuau, J. 2014. *Being Mizo*. Oxford: Oxford University Press.

Refugee Council USA. 2011. *Seeking Refuge: The Chin People in Mizoram State, India*. http://www.chinseekingrefuge.com (accessed 28 June 2019)

Sadiq, K. 2019. *Paper Citizens: How Illegal Immigrants Acquire Citizenship in Developing Countries*. Oxford: Oxford University Press.

Sollom, Richard, Adam K. Richards, Parveen Parmar, et al. 2011. 'Health and Human Rights in Chin State, Western Burma: A Population-Based Assessment Using Multistaged Household Cluster Sampling'. *PLoS Med* 8(2). https://doi.org/10.1371/journal.pmed.1001007 (accessed 9 August 2019).

Thang, Van Biak. 2011. 'Chin State Named Poorest in Burma'. *Chinland Guardian*, 7 July. http://chinlandguardian.com/news-2009/1406-chin-state-named-poorest-in-burma.html (accessed 6 August 2019).

The Telegraph. 2014. *Mizoram stares at Meth threat after haul*. The Telegraph, 7 May. https://www.telegraphindia.com/states/north-east/mizoram-stares-at-meth-threat-after-hauls/cid/185559 (accessed 30 September 2019).

The Transnational Institute. *About Ethnic Conflict in Burma*. https://www.tni.org/en/page/about-ethnic-conflict-burma (accessed 5 August 2019).

Times of India. 2017. *India opens two border crossing points with Myanmar*. Times of India, 1 October. https://timesofindia.indiatimes.com/india/india-opens-two-border-crossing-points-with-myanmar-bangladesh/articleshow/60899373.cms (accessed 30 October 2019).

UNHCR. 2014., *The Plight of the Myanmar's Chins in India*. 3 January. https://www.refworld.org/docid/52cbd70e4.html (accessed 30 September 2019).

UNHCR. 2019. *Analysis: The Plight of Myanmar's Chins in India.* https://www. unhcr.org.in/index.php?option=com_news&view=detail&id=8&Itemid=117 (accessed 7 August 2019).

UNOHCHR, 2019. *Myanmar: UN Expert Expresses Alarm at Escalating Conflict, Calls for Civilian Protection.* 18 January. https://www.ohchr.org/EN/ NewsEvents/Pages/DisplayNews.aspx?NewsID=24089&LangID=E (accessed 11 August 2019).

About the Editors and Contributors

EDITORS

H. Srikanth is a Professor in the Department of Political Science, North-Eastern Hill University (NEHU), Shillong, India. His research has primarily focused on issues concerning ethnicity, indigenous people and the political economy of Northeast India. Dr Srikanth was awarded the Shastri Indo–Canadian Faculty Research Fellowship to work on Native Indians in British Columbia. His publications include *Indigenous Peoples in Liberal Democratic States: Conflict and Reconciliation in Canada and India* (2010) and the co-edited volumes *Vision for Meghalaya: On and Beyond Inner Line Permit* (2014) and *Ethnicity and Political Economy in Northeast India* (2016). He has also published many research papers in such journals as *International Studies, Economic & Political Weekly* and *Man and Society*. During 2012–2017, he was the editor of the NEHU journal.

Munmun Majumdar is a Professor in the Department of Political Science, NEHU, Shillong, India. She completed her MA, MPhil and PhD at the School of International Relations, Jawaharlal Nehru University (JNU), Delhi. She was awarded the UGC Research Awards (earlier known as Career Award and National Fellow) for Post-Doctoral Research. She had been a Visiting Professor to University of Yangon, Myanmar, and Chulalongkorn University, Thailand. She has been a Visiting Fellow at Manohar Parrikar Institute for Defence Studies and Analyses and an Associate at Institute of Advance Study, Shimla. She has published three books and many research papers as book chapters and journal articles in journals such as *Strategic Analysis, India Quarterly, Indonesian Quarterly*, and *Asian Affairs*. Her areas of interest are Southeast Asian studies and international politics.

CONTRIBUTORS

Suparna Bhattacharjee has done her PhD from JNU and is working as an Assistant Professor in NEHU. Her specialization includes international political economy and East Asia. Some of her publications include: *Fencing the 'Commons': Battle for Right to Water Revisited through the Lens of 'Water Commons and Water Justice'* (2019); *Look East to Act East: Contextualizing India's Engagement with Singapore* (2018); and *Geography as an Opportunity: Border Haat in India's Foreign Policy Pursuit* (2019).

Rakhee Bhattacharya is an Associate Professor at Special Centre for the Study of North East India, JNU. She has been an Endeavour Post-Doctoral fellow in Australia. Her areas of research interest are political economy, development economics, regional economy, transnational economy and geo-economics, poverty and inequality, geopolitics, and India's Northeast and its neighbourhood. She has authored *Development Disparities in Northeast India* (2011) and *Northeastern India and its Neighbours: Negotiating Security and Development* (2015). Her latest edited volumes are *Regional Development and Public Policy Challenges in India* (2015) and *Developmentalism as Strategy: Interrogating Post-colonial Narratives on North East India* (2019).

Dechen Bhutia is a PhD scholar and a Senior Research Fellow at Centre for Inner Asian Studies, School of International Studies, JNU. Her areas of research interest include Central Asian and trans-Himalayan studies, international relations and the political economy of border regions. She is presently working on the topic 'Trans-Himalayan Trade through the Chumbi Valley of Tibet in the twentieth century'.

T. T. Haokip is a Professor in the Department of Political Science, NEHU, Shillong. He was a recipient of the South Asia Regional Fellowship from the Social Science Research Council (SSRC), New York, in 2005. He has completed two major research projects on Myanmar and Pakistan from SSRC, New York, and one major project on Democratic governance and Traditional Institutions of

North-East India from UGC. His current areas of interest include the interface between democratic governance and traditional institutions, ethnicity, ethnic conflicts and armed movements in Northeast India and South Asia.

Abu Hena Reza Hasan is a Professor in the Department of International Business, University of Dhaka. He has received postgraduate degrees in management, operations research and health policy. He has 27 publications, in academic journals and as book chapters, to his credit. Professor Hasan's research interest covers international trade, supply chain, healthcare management and agriculture. His present research interests are trade-related international agreements, analysis of airline failures and agricultural value chain.

Saurabh Kaushik is a PhD scholar in the Department of Peace and Conflict Studies and Management, School of Social Sciences, Sikkim University, Gangtok. He was the Director of Research & Development at Asian Institute of Diplomacy and International Affairs (AIDIA), Kathmandu, before pursuing his doctoral research. His areas of interest include strategic and security studies, conflict forecasting and analysis, conflict transformation and peacebuilding, non-traditional security threats primarily pertaining to South Asia and energy studies.

Mahendra P. Lama is a Professor of South Asian Economies at JNU, New Delhi. He was the founding Vice-Chancellor of Sikkim Central University and also served as Chief Economic Adviser to the Chief Minister of Sikkim during 2002–2009. He has been a recipient of prestigious fellowships and has been invited as a Visiting Professor in several reputed universities in India and abroad. Professor Lama is associated with several national and international organizations and has contributed immensely to the understanding of international politics and policy studies. Some of his recent works include: *North East Region Vision 2035* (2021); *Energising Connectivity between Northeast India and Its Neighbours* (2019); *Globalisation and Cultural Practices in Mountain Areas: Dynamics, Dimensions, and Implications* (2012); and *Human Security in India: Discourse, Practices, and Policy Implications* (2010).

P. C. Lalthansiami has completed her MPhil and PhD from NEHU, Shillong. She has worked on 'Internally Displaced Persons from Mizoram: A Study of the Brus' for her MPhil and 'Transborder Migration: A Study of Chin Migrants in Mizoram' for her doctoral degree. She has participated in different seminars and published a couple of journal articles.

Tejimala Gurung Nag is a Professor in the Department of History, NEHU, Shillong. Her field of specialization is modern history of India/Northeast India. Her thrust areas of research include Nepali migration and settlement in Northeast India, Christianity and colonial modernity and army and society in colonial Northeast India, on which she has published a number of papers. She has edited the book *The Falling Polities: Crisis and Decline of States in North-East India in the Eighteenth Century*. She was co-editor of the volume *Making of the Indian Union: Merger of Princely States and Excluded Areas*.

Sayada Jannatun Naim is Assistant Director of the Executive Development Center (EDC) of BRAC Institute of Governance and Development (BIGD), BRAC University, Dhaka, Bangladesh. Ms Naim has an MBA in international business and MSc and BSc (Honours) in child development and social relationship from University of Dhaka. She has four scientific publications related to international trade, governance and entrepreneurship. The title of her latest publication is 'Import of Health Care Services from India: Is it complementary to or Substitute of the National Health Service of Bangladesh?'

Bishnu Dev Pant is a professionally trained statistician, with over 35 years of research and managerial experience at both national and international levels. He served at the Asian Development Bank in different capacities and worked at the United Nations Economic and Social Commission for Asia and the Pacific (UNESCAP) in Bangkok as Chief of Statistical Information Services Section. He also served as the Executive Director of Institute for Integrated Development Studies (IIDS) in Kathmandu till 2019. He is presently teaching at the South Asian Institute of Management in Kathmandu. Dr Pant has a number of publications to his credit, including journal articles, to his credit in different areas.

Rakhal Kumar Purkayastha is a former Head and an Associate Professor in St. Anthony's College, Shillong. He received PhD from NEHU. He was Fulbright Scholar in residence during 2009–2010. He completed a Ford Foundation-supported project on Social Conflict: A Case Study of Tribal Non-Tribal Conflict (1981–2001). He has co-edited two books: *Human Disabilities—Challenges for Their Rehabilitation* (1998) and *Border Trade: India and Her Neighboring Countries* (2000).

R. K. Satapathy worked as a Professor in the Department of Political Science, NEHU. He is a former Honorary Director of Indian Council of Social Science Research's (ICSSR) North Eastern Regional Centre (NERC) and member of Lok Niti and was involved in election studies in the north-eastern states. His areas of interest are contemporary political theory, electoral politics and international relations. He was the editor of the ICSSR-NERC journal *Man and Society*. He has submitted a University Grants Commission (UGC)-sponsored Major Research Project entitled 'General Elections 2009—A Behavioural Study of Political Attitudes and Opinion in Meghalaya'. He authored *United States and Central America* (2012) and edited the volume *Globalisation and Traditional Systems in North East India* (2014).

Amrendra Kumar Thakur is a Professor and former Head, Department of History, NEHU, Shillong. He was the General Secretary of North East India History Association from 2007 to 2019. Professor Thakur has done extensive work on the history of Northeast India and published multiple writings on peasantry, slavery, economy, state formation, technology, colonial ethnography and trade. He has authored five books and edited several volumes. His important publications include: *India and the Afghans: A Study of a Neglected Region (1370–1576 A.D.)* (1992); *Slavery in Arunachal Pradesh* (2003); *History of Arunachal Pradesh (Early Times–1972 A. D.)* (2014); *Pre-Colonial Arunachal Pradesh* (2016); and *Technology of the Tribes of Northeast India* (2017).

Emdorini Thangkhiew is an Assistant Professor in Synod College in the Department of Political Science (PG Section). She has completed her MA and MPhil from the School of International Studies, JNU. Her areas of interest are world politics, international diplomacy and security studies. Her PhD research is on the refugee crisis in Europe. She is the editor of *Political Anvil*, the annual journal of the college.

Yedzin W. Tobgay taught political history at Royal Thimphu College in Bhutan. She studied political science and intelligence studies at Mercyhurst University and became the first Bhutan King's Scholar at University of Cambridge, where she did her MPhil in Modern South Asian Studies. Her dissertation included a comparative historical analysis of the role of Nepali and Bhutanese monarchies in the transition to democracy. Her research interests are gender, policy and Bhutanese history. She is currently an associate researcher with Phuensum Consulting Services.

Jelle J. P. Wouters is a social anthropologist and teaches in the Department of Social Sciences at Royal Thimphu College, Bhutan. He is the author of *In the Shadows of Naga Insurgency: Tribes, State and Violence in Northeast India* (2018) and *Nagas as a Society Against Voting and Other Essays* (2019) and the co-editor of *Nagas in the 21st Century* (2017) and *Democracy in Nagaland: Tribes, Traditions, and Tensions* (2018).

Hu Xiaowen is an Associate Professor in Institute of Indian Studies, Yunnan University. She worked in Yunnan Academy of Social Sciences from 2010 to 2018. She got her PhD degree from the Centre for East Asian Studies of JNU in September 2017, and currently she is the China India Scholar-Leader Fellow at India China Institute of The New School, New York, United States. Her research focuses on Indian foreign policy, Indian think tanks, Bangladesh–China–India–Myanmar (BCIM) and China–India relations. She is the co-editor of the volume *One Belt One Road: China's Global Outreach*, 2017. Her works have been published in *Global Review, Academic Forum, Chinese Social Sciences Weekly, Southeast and South Asian Studies*, etc.

Index